Sustainable Development Goals Series

The **Sustainable Development Goals Series** is Springer Nature's inaugural cross-imprint book series that addresses and supports the United Nations' seventeen Sustainable Development Goals. The series fosters comprehensive research focused on these global targets and endeavours to address some of society's greatest grand challenges. The SDGs are inherently multidisciplinary, and they bring people working across different fields together and working towards a common goal. In this spirit, the Sustainable Development Goals series is the first at Springer Nature to publish books under both the Springer and Palgrave Macmillan imprints, bringing the strengths of our imprints together.

The Sustainable Development Goals Series is organized into eighteen subseries: one subseries based around each of the seventeen respective Sustainable Development Goals, and an eighteenth subseries, "Connecting the Goals", which serves as a home for volumes addressing multiple goals or studying the SDGs as a whole. Each subseries is guided by an expert Subseries Advisor with years or decades of experience studying and addressing core components of their respective Goal.

The SDG Series has a remit as broad as the SDGs themselves, and contributions are welcome from scientists, academics, policymakers, and researchers working in fields related to any of the seventeen goals. If you are interested in contributing a monograph or curated volume to the series, please contact the Publishers: Zachary Romano [Springer; zachary.romano@springer.com] and Rachael Ballard [Palgrave Macmillan; rachael.ballard@palgrave.com].

Khalaf Al'Abri · Elizabeth Buckner
Editors

Higher Education and Sustainable Development in the Arab Gulf States

Practical Approaches and Institutional Strategies

Springer

Editors
Khalaf Al'Abri
College of Education
Sultan Qaboos University
Seeb, Muscat, Oman

Elizabeth Buckner
Department of Leadership, Higher and Adult Education
University of Toronto
Toronto, ON, Canada

ISSN 2523-3084 ISSN 2523-3092 (electronic)
Sustainable Development Goals Series
ISBN 978-3-032-19558-6 ISBN 978-3-032-19559-3 (eBook)
https://doi.org/10.1007/978-3-032-19559-3

This Springer imprint is published by the registered company Springer Nature Switzerland AG
The registered company address is: Gewerbestrasse 11, 6330 Cham, Switzerland

Acknowledgements

The editors and contributors would like to acknowledge the support of the Gulf Research Centre (GRC) and the 2023 Gulf Research Meeting (GRM), held at the University of Cambridge (UK) for their support, which allowed us to meet in-person to discuss our research and deepen our understandings of sustainability in the region. Without their support, this edited volume would not have been possible.

Contents

1 **Higher Education and Sustainable Development in the GCC: An Overview** ... 1
Elizabeth Buckner, Karma Salloum, and Khalaf Al'Abri

Part I Adapting Higher Education for Sustainable Development: Innovative Curricular and Pedagogical Practices

2 **Emirati Students and Climate Change Adaptation: Green Activism in the UAE?** ... 19
Tilde Rosmer

3 **Universities as Living Laboratories in the Gulf Region** ... 29
Anna Grichting Solder

4 **Using Project-Based Learning to Nurture Sustainable Competencies Among University Students in Kuwait** ... 45
Sosil Somokian

5 **Defining Education for Sustainable Development in the GCC: The Need for More Holistic Perspectives and Approaches** ... 59
Hajar Mahfoodh and Hessa AlAtawi

6 **Faculty Members' Perceptions of Higher Education for Sustainable Development: A Study from the College of Education at Qatar University** ... 73
Mayamin Altae

7 **Connecting Sustainability and Global Citizenship in Universities in the Gulf Cooperation Council** ... 81
Wendy A. Booth

Part II Institutionalizing Higher Education for Sustainable Development: Institutional and National Policies

8 Enhancing Climate Action Through Higher Education: The Role of GCC Universities in Advancing Nationally Determined Contributions (NDCs) and SDG 13 95
Quosay A. Ahmed, Tarig Z. Ahmed, Mawahib E. Ahmed, and Asia Mohamed

9 Green Emissaries: The Climate Change Leadership of Branch Campuses in the Arab Gulf States 109
Seungah Sarah Lee and Gabriela Reynell

10 The Role of Research Centers in Supporting Higher Education and Sustainable Development: Evidence from Sultan Qaboos University 123
Omer Hamad

11 The SDGs and Universities: Promoting Innovation and "Living Together" in the UAE 135
Salem Alhammadi and Abdeljalil Akkari

12 Omani Universities' Engagement with the SDGs: Current Practices and Future Opportunities 149
Munira Khalifa Ali Al Wahaibi

13 Urban Sustainability Education and the "Othering" of the City: Exploring SDG 11 in the GCC Context 161
Abdellatif Qamhaieh

14 Conclusion: Sustainability and Transformation in Arabian Gulf Higher Education 181
Khalaf Al'Abri and Elizabeth Buckner

Editors and Contributors

About the Editors

Khalaf Al'Abri is Associate Professor of Education Systems and Policies at Sultan Qaboos University, Oman, where he also serves as Head of the Department of Educational Foundations and Administration. His research focuses on higher education policy, governance, sustainability, and internationalization in the Gulf region. He has led several national and international research projects on sustainable development, quality assurance, and global citizenship education. Dr. Al-Abri has published extensively on university rankings and higher education reform. He is President of the Gulf Comparative Education Society and collaborates with UNESCO and other international partners to promote education for sustainable development.

Elizabeth Buckner is Associate Professor of Higher Education at the Ontario Institute for Studies in Education (OISE) at the University of Toronto, where she also holds the Tier 2 Canada Research Chair in Higher Education for Global Sustainable Development. Her research examines how global trends affect higher education and one strand of her work focuses on how universities are integrating sustainable development into their missions and operations. Dr. Buckner is a major contributor to the research literatures in the field of comparative education and comparative higher education. Her sole-authored book on higher education in the Arab world *Degrees of Dignity: Arab Higher Education in the Global Era* (University of Toronto Press, 2022) was praised as "a tour de force of a work on higher education in the Arab world." She has also published in many of the top journals in sociology of education (*Sociology of Education* and *the British Journal of Sociology of Education*); comparative education (*Comparative Education Review, Comparative Education, International Journal of Educational Development*); and comparative higher education (*Higher Education, Minerva*). She holds a MA in Sociology and a Ph.D. in International and Comparative Education, both from Stanford University.

Contributors

Mawahib E. Ahmed Institute of Engineering Research and Material Technology, National Center for Research of Sudan, Khartoum, Sudan

Quosay A. Ahmed University of Khartoum, Khartoum, Sudan

Tarig Z. Ahmed University of Cambridge, Cambridge, UK

Abdeljalil Akkari University of Geneva, Geneva, Switzerland

Munira Khalifa Ali Al Wahaibi Sultan Qaboos University, Muscat, Oman

Khalaf Al'Abri Sultan Qaboos University, Muscat, Oman

Hessa AlAtawi University of Bahrain, Skhir, Bahrain

Salem Alhammadi University of Geneva, Geneva, Switzerland

Mayamin Altae Faculty of Education, The University of Buckingham, Buckingham, UK

Wendy A. Booth University of South Wales, Cardiff, UK

Elizabeth Buckner University of Toronto, Toronto, Canada

Anna Grichting Solder University of Vermont, Burlington, VT, USA

Omer Hamad Sultan Qaboos University, Muscat, Oman

Seungah Sarah Lee HEC Paris, Doha, Qatar

Hajar Mahfoodh University of Bahrain, Skhir, Bahrain

Asia Mohamed University of Khartoum, Khartoum, Sudan

Abdellatif Qamhaieh American University in Dubai/United Arab Emirates, Dubai, UAE

Gabriela Reynell University of California, Berkeley, USA

Tilde Rosmer College of Humanities and Social Science, Zayed University, Dubai, UAE

Karma Salloum University of Toronto, Toronto, Canada

Sosil Somokian Box Hill College of Kuwait, Kuwait City, Kuwait

1 Higher Education and Sustainable Development in the GCC: An Overview

Elizabeth Buckner, Karma Salloum, and Khalaf Al'Abri

Abstract

In this introductory chapter, we provide an overview of higher education and sustainable development in the Arab Gulf region to situate the chapters that follow. The chapter first situates this book within a growing cross-national literature on the role of universities in contributing to sustainable development, which identifies important roles that universities can play in meeting the Sustainable Development Goals (SDGs). It then points out that the GCC is a distinctive and underexamined context for higher education and sustainable development, and that countries of the GCC have numerous challenges with environmental and social sustainability. A key idea that we find is that universities are playing important and even leading roles in GCC societies through their work on higher education and sustainability. This often takes the form of experiential and project-based learning in courses and through interdisciplinary and collaborative initiatives, including research. At the same time, we identify common challenges for universities across the region that limit their contributions, including difficulty with resources, commitment, and implementation.

Keywords

Higher education · Universities · Sustainable development · Gulf cooperation council

1.1 Higher Education and Sustainable Development

Around the world, universities are increasingly recognized as having an important role to play in promoting sustainability by providing expertise in research and education and building awareness of sustainable development-related issues (McCowan 2020; El-Jardali et al. 2018). The recognition of sustainable development as an important priority is a relatively new development: prior to the creation of the SDGs in 2015, higher education (HE) institutions were not meaningfully integrated into the international development agenda on sustainability (Chankseliani and McCowan 2021). Though some universities did begin to make commitments to sustainability in the 1970s, most of their efforts were limited to campus greening initiatives or other environment and campus-focused activities that did not target universities' other key functions, like teaching, research, and service (Stein 2024). Universities' efforts intensified in the 1990s after the Talloires

E. Buckner (✉) · K. Salloum
University of Toronto, Toronto, Canada
e-mail: elizabeth.buckner@utoronto.ca

K. Salloum
e-mail: karmasalloum@gmail.com

K. Al'Abri
Sultan Qaboos University, Muscat, Oman
e-mail: kabri@squ.edu.om

K. Al'Abri and E. Buckner (eds.), *Higher Education and Sustainable Development in the Arab Gulf States*, Sustainable Development Goals Series, https://doi.org/10.1007/978-3-032-19559-3_1

Declaration, a ten-point call to action for higher education institutions to integrate sustainability into all aspects of their operations (Dmochowski et al. 2016). Yet, commitments to sustainable development were less common.

The adoption of the United Nations Sustainable Development Goals in 2015 ushered in a new era—the SDGs embraced universities' role in contributing to sustainable development. Scholars and practitioners alike argue that higher education institutions can drive society's transition toward sustainability through their ability to advance research on sustainability and educate future generations (Chankseliani and McCowan 2021). For example, in his categorization, McCowan (2019) identifies five modalities through which universities can integrate sustainability into their institutions, namely, education, public debate, knowledge production, service provision, and value embodiment. The first, education, involves both teaching sustainability and critical thinking and incorporating sustainability into courses and curricula across various disciplines. The second is public debate, whereby universities create space for civil society debates and knowledge sharing on sustainability topics. The third modality is knowledge production, such as research within the university and in participation with local communities. The fourth is service provision, which involves supporting local communities and training students to become professionals who provide quality services. Finally, the fifth is embodiment, which is the idea that universities must reflect the values of sustainable development that they promote, such as environmental sustainability and equality, within all their operations and structures.

Over the past decade, universities have increasingly framed their social impact and contributions in terms of global challenges generally, and the SDGs in particular (Stein 2024). Many have adopted sustainability as a new mandate and core mission, moving beyond mere greening initiatives to integrating activities that also address the economic and social aspects of sustainability across universities' various roles (Buckner and Zhang 2024; McCowan 2019). To demonstrate their commitment to sustainable development, universities are also increasingly choosing to participate in international rankings and assessments that measure the impact of their sustainability initiatives, such as the Times Higher Education Impact Rankings or QS Sustainability Rankings.

However, scholars have raised concerns about the authenticity and real impact of their efforts. One of the biggest critiques of universities' participation in sustainability activities and rankings is that they reflect a form of greenwashing—when organizations convey an image of environmental sustainability but continue to engage in unsustainable practices that help them increase profit and economic growth (Hayes 2024). As Buckner and Zhang (2024) explained it, engaging in sustainability efforts improves universities' reputations and global standing, which helps them gain legitimacy and distinction. In turn, universities are more easily able to please their stakeholders, build partnerships, and access funding (Lakhno 2024; Stein 2024). Therefore, scholars have argued that many universities' initiatives may be largely symbolic. McCowan (2020) reinforces this view, highlighting that universities themselves can be significant contributors to climate change and thus should not be viewed as neutral actors whose sustainability commitments will undoubtedly result in positive outcomes.

In response, scholars have emphasized the need for universities to develop holistic and comprehensive strategies to address sustainability issues. To achieve this, they argue that universities must make sustainability a core organizing principle. As an organizing principle, sustainability must be adopted as a core value and key component of universities' organizing culture and identity (Buckner and Zhang 2024). Thus, as Kelly (2009) argues, sustainability should be integrated into all parts of an institution—from its operations, curriculum and teaching, research, and community engagement—simultaneously. By doing so, sustainability efforts are not implemented as isolated activities but are rather integrated into the entire institution to create real, long-lasting positive impact (McCowan 2019). Accomplishing this typically requires an interdisciplinary approach that involves all stakeholders, including students and faculty, in universities' efforts to advance sustainable development (Dmochowski et al. 2016).

Although the trend toward higher education for sustainable development is a global trend, it is important to keep regional and national contexts in mind. What it means to support sustainability is context-dependent: sustainability concerns and priorities are shaped by their national and local geographies, economies, politics, and societies. While there is a growing body of research on higher education for sustainable development in different national contexts, there has been very little focus on the topic in the Arab Gulf region. Recognizing the need to better understand this underexamined region, this book examines how universities are promoting sustainability in the countries of the Arabian Gulf, a region where sustainable practices have often been neglected in the name of economic development and modernization. In the sections that follow, we provide an overview of the region and its higher education system and then turn to the organization and key themes of the book.

1.2 An Overview of the Gulf Cooperation Council

The Gulf Cooperation Council (GCC) was established in 1981 as a regional organization and political and economic alliance between Bahrain, Kuwait, Oman, Qatar, Saudi Arabia, and the United Arab Emirates (UAE). Located on the Arabian Peninsula in the Middle East, the six GCC states are monarchies with Muslim-majority populations that share historical, cultural, and economic similarities, as well as conservative social norms and politics.

Until the mid-twentieth century, the Gulf was primarily composed of traditional Bedouin and tribal communities governed by local sheikhs (Buckner 2022). Although the region saw early colonial outposts by the Portuguese, French, and Dutch, the British had the largest colonial influence, maintaining its presence and control over the Gulf from 1820 until 1971. Britain's primary interest in establishing control over the region was for trade and security purposes, engaging less in religious mission work or cultural hegemony as compared to in other Arab states (Buckner 2022). Britain officially gained control over the Gulf in 1820 after establishing treaties with the Qawasim, the tribal group controlling the strategic maritime routes at the entrance to the Gulf, and began dominating the Gulf rulers' foreign policy decisions and militaries (Allday 2014). Saudi Arabia and Oman, which were never formally colonized by the British in the same way as the other Gulf states, officially declared independence and the beginning of their modern state building in 1932 and 1951, respectively. After decades of direct European colonial rule, Kuwait achieved independence in 1961, and Bahrain, Qatar, and the UAE gained independence in 1971, marking the end of British governance in the region.

The discovery of oil in the Gulf in the 1930s sparked drastic social, cultural, and economic changes that led to the region's rapid development. The expansion of natural resource extraction sparked major economic growth across the GCC, allowing the states to become some of the wealthiest in the world (ESCWA 2025). According to a report by the UN's Economic and Social Commission for Western Asia (2025), the average per capita income in all the GCC countries was two times greater than the global average. At the same time, this rapid growth was made possible by foreign labor. With the Gulf states developing rapidly and becoming some of the wealthiest of the world, large waves of migrant laborers, mostly from South Asia, settled in the Gulf to fill labor shortages and support the countries' growth from sheikhdoms with small populations to large, wealthy nation states with heavily populated urban areas (Buckner 2022). As a result, about half of the Gulf's residents are foreigners, and, interestingly, Oman and Saudi Arabia are the only two countries in the region whose populations consist mostly of national citizens (De Bel-Air 2025; The Economist 2018). Despite the prevalence of foreigners across the Gulf and their major contribution to the region's economic development, they do not have access to the same benefits that national citizens do.

As the ruling families across the region began amassing significant wealth from the development of the oil industry and oil exports globally, their social contracts with citizens transformed. One of the most significant developments was the evolution of the Gulf states into rentier states, whereby the rulers used their massive oil revenues, instead of taxes, to distribute benefits and state services to their citizens in exchange for loyalty, political compliance, and stability (Buckner 2022; The Economist 2018). For instance, citizens of the Gulf states do not pay income tax, benefit from subsidized housing, and receive preferential treatment in labor markets (Buckner 2022). In contrast, many migrant workers, who make up the majority of the labor force in most Gulf states, reportedly work in unregulated positions with little protection for their rights (Robinson 2022). Robinson (2022) also reports that migrant workers in the region often face poor working conditions, low wages, and abuse from their employers under the *kafala* (visa sponsorship) system, which is applied in several Gulf states to foreigners working in various key sectors and leaves workers' legal status tied to their sponsors.

Though the Gulf states share these historical and demographic similarities, they differ in their population size and political influence. Home to over half of the GCC's total population, Saudi Arabia is the largest Gulf state with over 32 million residents, followed by the UAE, whose population lies at around 10.2 million (De Bel-Air 2025). The remaining GCC states are relatively much smaller, ranging from Oman, who has the third-largest population at around 4.7 million, to Bahrain, the smallest, with a population of approximately 1.5 million people (De Bel-Air 2025). As the largest and arguably most influential Gulf state, Saudi Arabia currently acts as the de facto leader of the GCC, yet Baabood (2023) argues that increasing geopolitical competition between Saudi Arabia and the UAE has threatened Saudi dominance in the region. Despite these tensions, the GCC continues in its aims to foster "coordination, integration, and interconnection" among its member states, based on their "shared characteristics," similar political systems founded on Islam, and "unity of purpose" (GCC 2025). The GCC has thus launched a number of successful integrative economic projects that continue to drive growth for the region, such as a Free Trade Area, customs union, the GCC Interconnection Grid, which unites the power grids of the GCC countries into a single network, and common market (Baabood 2023).

The economic development of the region was largely driven by the GCC states' oil and gas revenues, which facilitated large-scale industrialization and urbanization in each country. Al-Saidi et al. (2019) described this as an "aggressive" strategy of "Petro-Urbanism" (p. 548) which scholars have argued depends heavily on fossil fuel extraction (Zaidan et al. 2019). While this model continues to create major economic growth for the region, it is highly unsustainable. For instance, the GCC countries have become some of the world's highest carbon dioxide emitters, contributing about half of the total CO_2 emissions from all Arab nations (Hammad et al. 2019). The region also generates substantial amounts of waste, with all six countries consistently ranking among the top ten per capita waste producers worldwide (Umar and Umeokafor 2022). Azar and Raouf (2017) have also shown that the GCC countries are among the fifteen highest per capita energy consumers globally, with Qatar at the top of the list. Despite these issues, GCC governments have historically failed to prioritize environmental sustainability in favor of protecting their carbon-based economies (Zaidan et al. 2019).

This has come at an environmental cost. Rapid and large-scale urbanization, high energy consumption, and the inefficient use of resources have given rise to serious environmental challenges across the region (Al-Saidi et al. 2019; Hammad et al. 2019). Almheiri et al. (2024) agree, arguing that infrastructure development and natural resource extraction projects have exacerbated the urban heat island effect and created environmental damage, making the GCC more vulnerable to climate change-related disasters and ecosystem degradation. For example, they report that in recent years, Oman has faced several damaging cyclones, while Kuwait, Oman, and the UAE have recorded extreme temperatures over 50 °C. Additionally, both Saudi Arabia and the

UAE have become increasingly susceptible to flooding (Shafi 2024). Given the region's arid climate, low elevation, desert environments, and sparse vegetation, the GCC is already highly vulnerable to climate change-induced disasters, such as floods, droughts, cyclones, and heat waves (Almheiri et al. 2024). Coastal cities, where most of the GCC population lives, are especially at risk (Zaidan et al. 2019). The effects of their reliance on oil, in addition to the GCC's unsustainable consumption patterns, have thus prompted the need for a transition toward more sustainable economic models and development pathways.

The GCC states recognized this and have become more involved in the global sustainability agenda, joining multilateral sustainability agreements and committing to the United Nations' 2030 Agenda for Sustainable Development and SDGs (Al-Saidi et al. 2019; Hammad et al. 2019). As part of their commitment, GCC governments have introduced national development plans and visions that align with the SDGs and position them as leaders in advancing sustainability in the region (Al-Saidi et al. 2019). Most national plans include goals of economic diversification away from oil and gas, the development of renewable energy and green tech, and water-use efficiency (Hammad et al. 2019). To achieve these ambitions, governments have invested heavily in a variety of large-scale greening initiatives. Saudi Arabia and the UAE, for instance, have built smart and sustainable cities with extensive renewable energy infrastructure as part of their goals to advance eco-friendly urban development (Almheiri et al. 2024). Koch (2024) argues, however, that "extravagant" projects like these serve as spectacles used to project an image of modernity and progress in the GCC (p. 335). By promoting narratives of sustainability and funding megascale green projects, GCC governments have been able to secure the political and reputational benefits associated with environmental sustainability action, while simultaneously avoiding major shifts away from their oil economies and dominance in the global energy sector (Koch 2023, 2024). In fact, Koch (2024) claims that these initiatives would not have been possible without the GCC states' revenues from oil exports.

Nevertheless, despite their stated commitment to sustainability, GCC countries have made slow progress toward achieving the SDGs (Umar and Umeokafor 2022). The 2024 Sustainable Development Report, which presented the SDG Index and Dashboards for 167 countries worldwide, found that global progress toward achieving the SDGs has largely stalled since 2020 (Sachs et al. 2024). Among the countries analyzed by Sachs et al. (2024), the GCC states ranked in the lower half, with the following positions from highest to lowest: UAE (70), Oman (95), Qatar (102), Saudi Arabia (103), Kuwait (111), and Bahrain (113). This reflects the persistent challenges faced by the GCC in aligning their development models with sustainable practices. Across the region, most countries have faced the greatest difficulty in achieving SDG 2 (Zero Hunger), SDG 6 (Clean Water and Sanitation), SDG 12 (Responsible Consumption and Production), and SDG 13 (Climate Action). Additionally, Saudi Arabia and Oman have not made sufficient progress on SDG 5 (Gender Equality), while Kuwait has struggled to achieve SDG 8 (Decent Work and Economic Growth) (Umar and Umeokafor 2022). Abdalla et al. (2024) generally attribute this to limited and ineffective multistakeholder partnerships, disagreement on priorities at the local level, insufficient political commitment, and difficulty in establishing effective governance structures.

Given these challenges and stalled progress, achieving sustainability in the GCC and beyond will require the participation and collaboration of multiple stakeholders, not merely governments (Azar and Raouf 2017). Scholars have called on higher education institutions to take a leading role in supporting sustainability in the GCC; however, it remains unclear to what extent their commitments have been integrated into their values and identities or translated into holistic, institution-wide efforts.

1.3 Higher Education in the GCC

Higher education in the Gulf has transformed from Islamic schooling into larger sociocultural and political institutions that play important roles in the Gulf countries' state building, national development, and global engagement. Today, universities across the Gulf are closely tied to cultural, economic, and political power throughout the region, and they continue to align their strategies with their states' national vision plans. As part of this, they are increasingly being called on to support national efforts to meet the SDGs.

Until the mid-twentieth century, education in the Gulf took the form of traditional Islamic education, which involved studying the Qur'an under religious scholars in tribal communities or attending religious schools to learn Arabic reading and writing and basic mathematics in addition to Islam (Alhebsi et al. 2015). After gaining independence, the newly formed Gulf states began establishing university systems as part of their modernization, state building, and national development process, marking the start of higher education institutions in the region (Buckner 2022). Most universities in the region were public, acting as arms of the government with clear nation-serving mandates. For instance, many Gulf states offered free university education to elite groups to prepare them for employment in the public sector as a means of securing their allegiance toward the new states instead of traditional tribes (Buckner 2022). HE institutions were also used to train professionals to meet the growing private sector's demands after the oil boom (Davidson 2012).

With globalization and the emergence of a new global era, higher education in the Gulf underwent major development and reform to align with the demands of the globalized world. HE systems in the region moved away from prioritizing the education of a small group of elites for government employment to providing citizens with the knowledge and skills required for work in a variety of sectors that support national economic development in the competitive global economy (Buckner 2022). Lee (2021) refers to this as an effort to build a "knowledge-based economy" in the GCC. As part of this shift, Gulf states have used education to invest heavily in nationalizing their labor forces in order to rely less on migrant laborers (Buckner 2022).

However, Lee (2021) argues that in addition to building a knowledge economy, the expansion of HE in the region has also served to improve Gulf states' international legitimacy and reputation. For instance, many universities engage in accreditation and global rankings to help their states gain prominence in the international arena (Lee 2021). Education policies across the region also encourage openness to foreign partnerships, such as by inviting international universities to set up branches in the Gulf or adopting Western universities' models and best practices (Badry 2019). In fact, Koch (2018) argues that the Gulf's most prestigious universities and their curricula, programs, and missions were directly modeled on American higher education institutions. Universities in the Gulf also attract large numbers of foreign students, demonstrating the international, or regional, standards of their education (Umar and Rahman 2023).

Gulf universities' increased internationalization has, however, led to some backlash and the call for more traditional institutions that preserve and promote national, Arab culture and heritage (Badry 2019). As a result, according to Lee (2021), Gulf states today have two general types of HE institutions: private, international universities that follow Western models of education and are more globally oriented, and public universities that promote the states' national identity and cater more heavily to local and Arab populations to gain domestic legitimacy. The HE sector continues to grow in the GCC, with each state making commitments to improving HE in their national visions and strategies and increasingly allocating large shares of their budgets to education (Mosly 2022). Table 1.1 provides an overview of higher education systems in GCC countries.

Universities across the Gulf are increasingly being recognized for playing an important, potentially leading role in advancing the region's sustainable development. Scholars have described how many express commitment to combating sustainability-related issues and working toward the SDGs in their missions and visions, with some even describing themselves as "living laboratories of sustainability," projecting an image of themselves as spaces that embody sustainability across their institutions (Koch 2018, p. 526). However, the way that universities approach sustainable

Table 1.1 Overview of higher education systems in the GCC

Country	Year first university founded	# Public universities	# Private universities	Total higher education enrollments
Bahrain	1986	4	14	54,099
Kuwait	1966	2	14	121,840
Qatar	1977	13	20	46,283
Saudi Arabia	1957	29	38	1,572,022
Oman	1986	3	27	117,981
United Arab Emirates	1976	4	69	339,951

Notes Data refers to most recent year available. Public universities include security and military higher education institutions. Private universities include international branch campuses and all those in Qatar's Education City. *Sources* UNESCO Institute for Statistics (2025); Information and eGovernment Authority, Kingdom of Bahrain (n.d.); Kuwait Government Online (n.d.); Kuwait Education Information (n.d.); Abdullah Al Salem University (n.d.); International University of Kuwait (n.d.); Ministry of Education and Higher Education, State of Qatar (2024); Ministry of Education, Kingdom of Saudi Arabia (n.d.); Ministry of Higher Education, Research and Innovation, Sultanate of Oman (2022); Ministry of Higher Education and Scientific Research, United Arab Emirates (2023);Ibrahim and Barnawi (2023)

development in practice has been relatively unexplored. In this book, we discuss the ways in which universities across the GCC have been promoting sustainability and particularly highlight their work in teaching and learning to change students' mindsets and transfer the values of sustainability to future generations.

Universities in the region have introduced sustainability initiatives across their functional areas, but for most, their sustainability efforts have primarily been faculty-led projects involving campus greening, sustainability-related research, and integrating sustainability into teaching. For instance, Koch (2018) discusses GCC universities' "green laboratories," whereby campuses are transformed into more sustainable ones through "green features." These often include renewable energy, electric transportation, energy-efficient building infrastructure, recycling, and waste reduction projects (Al Mahameed et al. 2023; Koch 2018). GCC universities are also considered to have high research productivity in comparison to other Arab ones, contributing to advancing sustainability in the Middle East by investing heavily in grants and scholarships for sustainability-related research (Abdalla et al. 2024). Qatar University, for example, has been funding research projects that explore energy and the environment (Zaidan et al. 2023). These research initiatives are also strategically aligned with the GCC states' broader goal of building knowledge-based economies and societies (Alghamdi 2018). Furthermore, as part of their sustainability efforts, universities are also integrating sustainability as a theme into their teaching and developing sustainability-focused courses, which Albattah and Bande (2023) show have helped to increase students' awareness of environmental issues and encourage their participation in campus-based sustainability activities.

Though much is being done and most GCC universities continue to participate in sustainability ranking systems to measure their impact, some scholars debate the real impact of universities' efforts. Many researchers have highlighted that despite universities in the Gulf having introduced numerous sustainability initiatives, they usually only address environmental challenges, namely through high-tech solutions, or economic development, ignoring the other important aspects of sustainability (Alghamdi 2018). Particularly, studies have shown a major lack of attention to gender issues and other social barriers (Abdalla et al. 2024). Magd and Bystrov (2021) propose that this is due to GCC universities' need to align with national vision documents, which prioritize economic growth and development. Alghamdi (2018) also critically highlights that though sustainability goals are set by universities in the Gulf, they lack clearly defined plans and policies to achieve them. He also argues that most universities do not have sustainability offices or programs to train faculty and staff on sustainability issues, claiming that overall, not many universities have adopted holistic and institution-wide approaches to sustainability. In addition to these limitations, Gulf universities tend to push for highly visible sustainability efforts, such as the "living laboratories" concept, which involves large-scale and high-tech campus greening initiatives. Koch (2018) claims that these often serve more as campaigns to boost the universities', and thereby the Gulf nations', image and international legitimacy instead of meaningful sustainability initiatives. Moreover, she highlights that these types of technology-driven projects often distract from deeper environmental issues, prioritizing symbolic projects over institutional transformation.

These critiques have led to debates over the limits of universities' roles. Indeed, there are also debates over how to best institutionalize change and whether top-down or bottom-up approaches are most effective. In this book, the authors address many of these issues.

1.4 About the Book and Key Themes

This book explores and analyzes the role played by higher education institutions in the Gulf region in advancing sustainable development. It emerged from an in-person workshop entitled "The Leading Role of Gulf Higher Education in Achieving Sustainability and Addressing Climate Change," sponsored by the Gulf Research Meeting held at the University of Cambridge, England in July 2023. At the workshop, participants shared preliminary research findings, provided feedback and engaged in collective discussion of shared experiences. Throughout the discussions, it was apparent that universities in the region are responding to the call to integrate sustainability throughout their organizations and that many of these changes are being institutionalized into formal policies and practices. At the same time, we found common challenges, some of which were distinctive to the region. Through these discussions, we organized the contributions into two sections, broadly characterized as adaptation and institutionalization.

The first section, *Part I—Adapting Higher Education for Sustainable Development: Innovative Curricular and Pedagogical Practices* focuses on classroom or curriculum-based interventions and activities, including innovative teaching models such as problem-based learning and Living Labs, and their potential to engage students and advance sustainability. It also highlights the role of faculty and student perceptions of sustainability. The chapters in *Part II—Institutionalizing Higher Education for Sustainable Development: Institutional and National Policies* examine HESD as a national and institutional project, highlighting universities' efforts to enact structural change, introduce institution-wide initiatives, and align their strategies and visions with national sustainability goals and projects. These include the establishment of research centers, strategic frameworks, sustainability policies, and monitoring and reporting mechanisms.

Collectively, the chapters present various perspectives on how universities across the Gulf states have engaged in sustainability practices and highlight several practical, real-world applications of sustainability initiatives ranging from the teaching to institutional level. Most of the chapters in this book particularly highlight universities' efforts to teach students about sustainable development challenges and transfer to them the values of sustainability, as well as their work advancing research on sustainability-related topics—mainly climate change. The contributors also discuss the challenges that universities in the Gulf face and reveal some key institutional and structural limitations to their roles in driving sustainable development in the region.

Across the chapters, several key cross-cutting themes emerged. First, contributors point out that universities are in many ways, already leaders and innovators in advancing sustainability, with many playing significant roles in promoting climate literacy, supporting sustainability-oriented research, and equipping students with the skills needed to address environmental and social challenges. Second, the chapters highlight the need for interdisciplinary approaches and the need to localize sustainability education by aligning it with Gulf societies' cultural values and diverse social contexts, thereby enhancing students' engagement and the relevance of sustainability curricula. At the same time, the chapters identify numerous challenges that universities in the region are facing with implementation and institutionalization. Among these, key challenges included limited funding, fragmented institutional strategies, and insufficient faculty training and awareness. Additionally, in terms of implementation, contributors point out that most sustainability initiatives in the GCC have been shaped by government agendas and institutional leaders, aligning with national visions and global frameworks. While such top-down efforts can provide coherence and resources, several authors argue that bottom-up engagement, driven by students, faculty, and local communities, is essential for cultivating a genuine sustainability culture. Finally, authors point out that despite growing interest, most institutions lack standardized frameworks for evaluating the impact of their work. The book's contributors advocate for more transparent reporting mechanisms and for expanding the reach of sustainability activities to include diverse and marginalized communities, ensuring that universities' contributions to sustainable development are both meaningful and inclusive. In what follows, we elaborate on each of these key themes.

1.4.1 Universities as Leaders and Innovators for HESD in the GCC

A key idea from contributors is that universities in the GCC are already playing important, if not leading, roles in driving sustainable development in the Gulf, and they should continue to do so. In their chapter, Awad et al. (Chap. 8) focus on climate-related programs offered by Gulf universities and argue that they can help universities drive transformative, sustainable change in their communities by promoting climate literacy among their students, supporting sustainability-focused research projects, and equipping the future generations with the knowledge they need to tackle sustainability issues. In their chapter, Lee and Reynell (Chap. 9) agree, claiming that universities in the GCC have already been portrayed as hubs for green technology and sustainability. In analyzing the sustainability narratives and initiatives of American universities' branch campuses in Qatar and Abu Dhabi, Lee and Reynell also highlight that while supporting the sustainability goals of their host countries, branch campuses have been active participants in the Gulf states' sustainability efforts as well, even hosting national sustainability conferences and events. This "dual embeddedness" has allowed branch campuses in the Gulf to position themselves as both local and global leaders in sustainable development.

With regard to curriculum and teaching-based change, the book's contributors highlight the value of hands-on learning, project-based, and problem-based approaches to teaching sustainability in the GCC. In his exploration of an architecture course offered by the American University in Dubai that incorporates hands-on learning experiences, Qamhaieh (Chap. 13) argues that teaching sustainability in an abstract and theoretical manner is not effective for helping students gain a full understanding of sustainability. Rather, Qamhaieh advocates for linking theory to practice and pushes for the introduction of ESD courses that provide students with the opportunity to witness, experience, and contribute to solving sustainability-related issues firsthand—through project-based exercises, for instance. Qamhaieh's chapter shows that these types of learning experiences can help students develop strong problem-solving, critical thinking, and reflective thinking skills which are essential for thinking about and addressing sustainable development. Rosmer also highlights the transformative potential of experiential learning and project-based courses for students to increase their engagement with sustainability.

Somokian (Chap. 4) illustrates the transformative potential of such courses by examining three project-based learning courses offered at the Box Hill College Kuwait. As part of these courses, students were exposed to real-world environmental issues and asked to develop practical solutions, including awareness campaigns for example. Somokian (Chap. 4) finds that those enrolled in these project-based courses not only gained an increased knowledge of sustainability and the environmental challenges affecting their country, but they also fostered a sense of responsibility for addressing sustainability issues. In her research in Qatar, Altae (Chap. 6) agrees with Somokian's argument and highlights that curriculum and program-based changes like these can really drive students' engagement and understanding of how to advance sustainable development. As part of this, she suggests that universities integrate internship and research opportunities into their programs to help students gain meaningful practical experience. Grichting (Chap. 3) also explains how Living Labs are also sites of experiential learning where students can gain direct exposure to sustainable technologies, agricultural practices, and topics like biodiversity preservation. By participating in their campuses' community gardens or food production initiatives, or getting involved in planning and designing sustainable spaces, university students can acquire greater practical experience to complement their theoretical learning.

1.4.2 The Need for Interdisciplinary and Contextualized Approaches to HESD

The chapters also emphasize that the complexity and multidimensionality of sustainability challenges require interdisciplinary approaches for higher education for sustainable development. Several contributors observe that universities across the region tend to incorporate sustainability selectively—limiting the integration of ESD to specific disciplines rather than across the entire institution's departments. Somokian (Chap. 4), however, emphasizes that because sustainability issues touch on multiple areas, successful HESD requires universities to take lessons from various disciplines to equip students with diverse skills. Booth (Chap. 7) builds on these arguments in her chapter where she encourages educators to use methods from other disciplines to improve their teaching strategies. For instance, Booth suggests the use of concepts from social psychology and sociology to help students foster greater understanding of others, reduce prejudice, support inclusive attitudes, and foster a global citizenship mindset, which she argues is essential for advancing sustainable development.

Another key idea is the need for universities to engage beyond academic groups and collaborate with various sectors of society—from the local to the international levels. Contributors like Grichting (Chap. 3) and Mahfoodh and Al Atawi (Chap. 5) advocate for cooperation between universities and governments, private sector, international organizations, civil society, and local communities to exchange knowledge, create practical learning opportunities for students, access technical expertise and global sustainability frameworks, and, importantly, ensure that universities' initiatives to support sustainable development are relevant.

Similarly, several authors highlight the importance of adapting sustainability initiatives and strategies to the local culture and communities' needs. Booth (Chap. 7) argues that Western models for sustainable development are often ineffective in the Gulf and must be localized to align with local values, cultures, and perspectives. Rosmer (Chap. 2) supports this argument, noting that the lack of culturally and geographically contextualized approaches to higher education for sustainable development in the UAE contributes to a lack of student engagement with the content, as they find it difficult to relate to the material. Qamhaieh (Chap. 13) stresses that teaching sustainability should involve encouraging students to "invest in what they already have," thereby localizing the concept and making it relatable, which he argues would make it more interesting for students. Importantly, however, Alhammadi and Akkari (Chap. 11) discuss that many countries in the Gulf, like the UAE, have highly diverse populations; thus, universities' sustainability content and initiatives should reflect the multicultural local context as well.

1.4.3 Challenges to Institutionalizing HESD in the GCC

Despite a shared recognition of universities' importance in supporting sustainability efforts, several authors acknowledge that there are significant institutional and structural limitations that constrain HEIs' impact on sustainable development. One issue is funding gaps. In his analysis of the Sultan Qaboos University in Oman, Hamad (Chap. 10) examines the extent to which the university's research centers have undertaken meaningful climate change research projects but finds that since the centers' work aligns with state priorities, the government's lack of emphasis on climate change research has resulted in little funding for this type of research. Hamad argues that limited funding, in addition to absence of institutional strategies and a coordinated research agenda, has led to fragmented and insufficient climate change research efforts.

Another issue highlighted across the chapters is the fact that many universities in the GCC face gaps between their sustainability goals and actual outcomes. Mahfoodh and Al Atawi (Chap. 5) for instance, argue that despite an intention to do otherwise, universities' sustainability efforts often take the form of isolated initiatives, such as casual conferences and forums, instead of effective systematic approaches to integrating education for sustainable development into their entire curricula. Similarly, in the context of Omani universities, Al Wahaibi (Chap. 12) discusses that while universities often plan to make progress towards the SDGs on paper, their impact is often limited due to persistent gaps between institutions' intentions and actual outcomes. Hamad asserts that this limited effective contribution to sustainable development has occurred due to the lack of specific, university-wide frameworks guiding the institutions' sustainability efforts. For example, instead of being driven by an institution-wide agenda that promotes greater sustainability and climate change research, Hamad argues that the climate projects that do exist at the Sultan Qaboos University are largely the result of individual research center directors' personal initiative.

In addition to institutional-level challenges and implementation gaps, contributors highlighted educators' lack of awareness of sustainable development as a challenge to integrating sustainability into teaching and curricula. In her chapter, Altae (Chap. 6) explores the potential of faculty members at Qatar University's College of Education to drive education for sustainable development by educating Qatar's future teachers on climate change and other sustainability issues. However, the chapter argues that the College's educators have limited understandings of sustainability, which affects their ability to effectively integrate related topics into their teaching and transfer sustainability values and knowledge to their students. Al Wahaibi (Chap. 12) highlights a similar case in her chapter and shows that faculty members at Omani universities lack awareness of sustainability issues and frameworks. Thus, contributors like Booth (Chap. 7) and Mahfoodh and AlAtawi (Chap. 5) both stress the need for greater sustainability training for university educators, including on areas like social justice and global citizenship which Booth argues are essential for advancing sustainable development.

Throughout the volume, the chapters show how these institutional and teaching-related challenges can be intensified by ambiguities over what sustainable development actually entails. In fact, a common phenomenon viewed by authors like Qamhaieh (Chap. 13) and Mahfoodh and AlAtawi (Chap. 5) is that education for sustainable development is being widely misinterpreted by some universities in the region. Mahfoodh and AlAtawi (Chap. 5), for instance, argue that education for sustainable development is often simply identified with teaching students about climate change and environmental issues, and that sustainability initiatives are reduced to what Qamhaieh (Chap. 13) terms "technological solutions." Booth (Chap. 7) reinforces these claims in her chapter, where she explores the challenges of promoting education for sustainable development within GCC universities. She argues that a narrow concentration on the environmental aspects of sustainability limits universities' ability to implement broad, holistic sustainability initiatives. She stresses that universities should also integrate the economic and social dimensions of sustainability into their sustainability goals and activities to strengthen their education for sustainable development efforts. She also goes beyond the other authors and advocates for global citizenship and social justice concepts to be included within education for sustainable development frameworks as well.

1.4.4 The Need for Bottom-Up Initiatives

One of the major debates that emerged in the chapters on implementation concerned whether institutionalization should occur through top-down or bottom-up initiatives. As the chapters show, GCC universities' sustainability efforts have mostly been introduced due to top-down directives, whether because of government-determined priorities and encouragement or decisions made by the universities' leadership, ultimately in response to global sustainability agendas.

A major theme identified across the chapters is GCC universities' alignment with their governments' national priorities and strategies. Al Wahaibi (Chap. 12) shows that in Oman, universities' sustainability planning is a top-down effort, whereby leaders introduce plans that are heavily linked to the state's national vision and sustainability goals. This reinforces Hamad's observation that sustainability research conducted by the Sultan Qaboos University is largely guided by the Omani government's defined priority areas. As Lee and Reynell (Chap. 9) argue, even American university branch campuses in the GCC implement sustainability initiatives that not only support but are closely coordinated with the sustainability efforts and activities of their host countries. The chapters also show that international institutions, such as UNESCO, and agreements like the Talloires Declaration orient Gulf universities' sustainability strategies toward global agendas.

Many authors assert that such top-down initiatives are important for driving institutional change and call for various governance changes to improve the integration of sustainability into universities. Mahfoodh and AlAtawi (Chap. 5), for instance, recommend the establishment of high-level units focused on planning for the implementation of education for sustainable development and sustainability activities across the university. Similarly, in their analysis of climate-related programs offered by universities across the GCC, Awad et al. (Chap. 8) advocate for greater institutional coordination and centralized resources for sustainability education. Moreover, Hamad (Chap. 10) argues that in order to advance climate change research, universities need to include climate change as a priority in their wider institutional agendas. Booth (Chap. 7) also reaffirms the need for universities to establish clear, institution-wide frameworks that integrate all aspects of sustainability and the SDGs into research, teaching, and operations for more effective sustainability efforts. Finally, Rosmer (Chap. 2) agrees that top-down initiatives like these are already more prevalent throughout the Gulf, especially as there has scarcely been any strong push for sustainability action from students in the GCC.

Indeed, several chapters find that students in the GCC are more commonly passive recipients of sustainability activities rather than active participants in shaping sustainability plans and priorities. Rosmer (Chap. 2) evaluates how Emirati university students respond to their leaders' climate change discourses by analyzing students' environmental knowledge and engagement. She finds that there is a general lack of environmental knowledge, engagement, and activism among students in the UAE. Though Emirati students recognize that climate change is an issue, Rosmer (Chap. 2) argues that they have a low level of knowledge about its causes and effects, especially in their country and the wider region. Likewise, Qamhaieh (Chap. 13) explains that most students in the GCC are disconnected from sustainability issues as they have been conditioned to live "a life on the periphery," observing changes around them with little engagement or sense of responsibility over sustainable development. Mahfoodh and AlAtawi (Chap. 5) make similar observations, pointing to a lack of what they call sustainability culture.

Collectively, the book's chapters suggest that students' limited engagement with sustainability remains a major barrier to bottom-up initiatives, and as many contributors emphasize, effective and sustainable change within university communities requires greater student engagement and bottom-up participation. In their chapters, Somokian, Alhammadi and Akkari, and Rosmer all argue that higher education for sustainable development should instill a sense of responsibility in students for sustainability and their local environment, empowering them to act as agents of change. Grichting (Chap. 3) illustrates the importance of such engagement by presenting a successful case of bottom-up and collaborative action at Qatar University's Living Laboratory. She describes how the Lab, envisioned by researchers and faculty at the university, was developed and implemented collaboratively, involving non-governmental organizations, civil society, university staff, and students, to advance education on sustainable development at the university and address the present-day sustainability challenges in the Gulf. This case illustrates the value of multistakeholder and student engagement in policymaking and project design.

Many chapters, especially those in Part One, present strong arguments that advancing sustainability can be achieved through initiatives that incorporate experiential learning approaches, as they can increase students' engagement and help them play a greater role in higher education for sustainable development.

1.4.5 Calls to Better Assess the Impact of HEIs

The question of how to better assess the impact of universities' sustainability initiatives was another recurring debate across chapters. Hamad (Chap. 10) argues that more research is needed to evaluate the tangible benefits of climate research conducted by universities, reflecting broader gaps in understanding institutional impact. Similarly, Al Wahaibi notes the limitations of current impact assessment tools, arguing that aside from reviewing universities' sustainability reports and comparing them to the institution's goals, there have been limited ways to effectively measure their impact. A few authors argue that some institutions, however, have developed robust internal assessment mechanisms for tracking their sustainability progress. For instance, Lee and Reynell (Chap. 9) note that Georgetown University Qatar uses an in-house sustainability dashboard to collect and report relevant data, and NYU Abu Dhabi has developed a methodology for calculating its greenhouse gas emissions to include in reporting as well. Despite cases like these, in their exploration of four UAE universities' efforts to integrate education for sustainable development into their institutional frameworks and report on their activities, Alhammadi and Al Akkari (Chap. 11) argue that universities across the region still need stronger reporting and visibility on their contributions to the SDGs. They call for a clear and standardized sustainability reporting framework and comprehensive strategies to address all the SDGs.

In addition to debates about how universities assess and report their impact, the book also highlights the importance of critically examining who universities are reaching. Specifically, several contributors have raised concerns about the ways in which universities may inadvertently reproduce exclusivity in their sustainability efforts. Lee and Reynell (Chap. 9) highlight how universities in the GCC may be using sustainability as a tool to portray themselves as global leaders in higher education for sustainable development and increase their reputation and legitimacy potentially without addressing broader inclusion or equity issues. Qamhaieh (Chap. 13) also reminds us that many universities in the GCC are elite spaces that can serve only a small segment of the population. Thus, he emphasizes the need for sustainability education to broaden students' perspectives and reach. One way of doing so is by encouraging students to work and engage with local community organizations, especially with those based in rural areas, to ensure that universities' sustainability efforts are not concentrated among urban and elite populations or reinforcing structural exclusion.

1.5 Concluding Remarks

In conclusion, this book offers a critical understanding and analysis of the role that higher education systems in the Arab Gulf states play in global and national efforts to advance sustainable development. Given that the Gulf states are the largest producers of oil and gas in the world, the book presents unique insights on how higher education institutions across the region are actively supporting their governments' transition toward creating more sustainable countries. Contributing to the literature on higher education and sustainable development more broadly, the book presents insights into how universities in the GCC are integrating sustainability into their institutional frameworks, strategies, and curricula. In addition to synthesizing literature on higher education and sustainability in the region, many chapters also draw on primary data, including surveys and interviews with university administrators, faculty, and students. Thus, the book goes beyond presenting theoretical concepts and showcases universities' practical approaches to advancing sustainability that can be used to inform practice in other higher education contexts in the wider MENA region, Global South, and worldwide. It also highlights the challenges and lessons learned by universities in the region and offers actionable recommendations to inform regional and national policymaking and improve institutional efforts. We hope the insights from this book will engage scholars, researchers, policymakers, and practitioners in broader discussions on higher education and sustainability and inspire greater cross-disciplinary collaboration to effectively integrate sustainability across the higher education sector.

References

Abdalla S, Ramadan E, Al-Belushi MAK, Al-Hooti N (2024) Unveiling the role of Arab universities in advancing Sustainable Development Goals: a multidimensional analysis. Sustainability 16(14):5829. https://doi.org/10.3390/su16145829

Abdullah Al Salem University (n.d.) About Abdullah Al Salem University. Abdullah Al Salem University. https://www.aasu.edu.kw/

Al Mahameed M, Riaz U, Aldoob MS, Halari A (2023) The implementation of sustainability practices in Arab higher education institutions. J Finan Report Account 23(3):1161–1185. https://doi.org/10.1108/JFRA-11-2022-0415

Albattah M, Bande L (2023) Awareness and perception of the environmental sustainability of the UAE University campus: a case study of sustainability course. Int J Sustain Higher Educ 24(7):1610–1628. https://doi.org/10.1108/IJSHE-04-2022-0129

Alghamdi N (2018) Calm before the storm: assessing climate change and sustainability in Saudi Arabian Universities. In: Filho WL, Manolas E, Azul AM, Azeiteiro UM, McGhie H (eds) Handbook of climate change communication: practice of climate change communication, vol 2. Springer, pp 317–340

Alhebsi A, Pettaway LD, Waller L (2015) A history of education in the United Arab Emirates and Trucial sheikdoms. Global eLearn J 4(1):1–6

Allday L (2014) The British in the Gulf: an overview. Qatar National Library. https://www.qdl.qa/en/british-gulf-overview

Almheiri A, Montenegro JF, Ewane EB, Mohan M (2024) Climate change hazards and the resilience of coastal cities in the Gulf Cooperation Council countries: a systematic review. City Environ Interact 24:100177. https://doi.org/10.1016/j.cacint.2024.100177

Al-Saidi M, Zaidan E, Hammad S (2019) Participation modes and diplomacy of Gulf Cooperation Council (GCC) countries towards the global sustainability agenda. Develop Pract 29(5):545–558. https://doi.org/10.1080/09614524.2019.1597017

Azar E, Raouf MA (2017) Outlook towards the future of sustainability in the gulf. Sustainability in the gulf. Routledge, pp 277–280

Baabood A (2023) The future of the Gulf Cooperation Council Amid Saudi-Emirati Rivalry. Carnegie Endowment for International Peace. https://carnegieendowment.org/research/2023/10/the-future-of-the-gulf-cooperation-council-amid-saudi-emirati-rivalry?lang=en

Badry F (2019) Internationalization of higher education in the countries of the Gulf Cooperation Council: impact on the national language. In: Background paper prepared for the Arab States 2019 global education monitoring report: migration, displacement and education: building bridges, not walls. UNESCO. https://unesdoc.unesco.org/ark:/48223/pf0000371703

Buckner E, Zhang Y (2024) Sustainability as an emerging mandate in higher education. Comp J Compar Int Educ 14:1–19. https://doi.org/10.1080/03057925.2024.2429819
Buckner E (2022) Degrees of dignity: Arab higher education in the global era. University of Toronto Press. https://directory.doabooks.org/handle/20.500.12854/115898
Chankseliani M, McCowan T (2021) Higher education and the sustainable development goals. Higher Educ 81:1–8. https://doi.org/10.1007/s10734-020-00652-w
Davidson C (2012) Higher education in the Gulf: a historical background. In: Davidson C, Smith PM (eds) Higher education in the Gulf states: shaping economies, politics and culture. Saqi
De Bel-Air F (2025) National and foreign populations in GCC countries (Factsheet No. 13). Gulf Labour Markets and Migration. https://gulfmigration.grc.net/wp-content/uploads/2025/02/Francoise-De-Bel-Air-Factsheet-No.-13-GCC-Populations-2025-02-27.pdf
Dmochowski JE, Garofalo D, Fisher S, Greene A, Gambogi D (2016) Integrating sustainability across the university curriculum. Int J Sustain Higher Educ 17(5):652–670. https://doi.org/10.1108/IJSHE-10-2014-0154
El-Jardali F, Ataya N, Fadlallah R (2018) Changing roles of universities in the era of SDGs: rising up to the global challenge through institutionalising partnerships with governments and communities. Health Res Policy Syst 16(38):1–5. https://doi.org/10.1186/s12961-018-0318-9
ESCWA (2025) Real sizes of Arab economies between 2017 and 2023. United Nations Economic and Social Commission for Western Asia. https://icp.unescwa.org/charts/ICP-Report.pdf
Gulf Cooperative Council (GCC) (2025) About us - primary law. https://www.gcc-sg.org/en/AboutUs/Pages/PrimaryLaw.aspx
Hammad SH, Al-Saidi M, Zaidan E (2019) Beyond the pledges: reflections on sustainability transitions in the Gulf Cooperation Council (GCC) region. Develop Pract 29(5):539–544. https://doi.org/10.1080/09614524.2019.1628920
Hayes A (2024) Greenwashing: definition, how it works, examples, and statistics. Investopedia. https://www.investopedia.com/terms/g/greenwashing.asp
Ibrahim A, Barnawi O (2023) The past, present, and future of higher education in the Arabian Gulf region: critical comparative perspectives in a neoliberal era. Routledge
Information and eGovernment Authority, Kingdom of Bahrain (n.d.) Number of higher education institutions by sector and type of institution. https://www.data.gov.bh/explore/embed/dataset/21-hec_higher-education-institutes-by-sector/table/?disjunctive.sector&disjunctive.type_of_institution&disjunctive.year&sort=type_of_institution&refine.sector=Private&refine.sector=Government
International University of Kuwait (n.d.) Vision, mission and values. International University of Kuwait. https://www.iuk.edu.kw/
Kelly T (2009) Sustainability as an organizing principle for higher education. In: Aber J, Kelly T, and Mallory B (eds) The sustainable learning community: one university's journey to the future. University of New Hampshire Press, pp 1–53
Koch N (2018) Green laboratories: university campuses as sustainability "exemplars" in the Arabian Peninsula. Soc Nat Resour 31(5):525–540. https://doi.org/10.1080/08941920.2017.1383546
Koch N (2023) Sustainability spectacle and 'post-oil' greening initiatives. Environ Polit 32(4):708–731. https://doi.org/10.1080/09644016.2022.2127481
Koch N (2024) Sustainability spectacle in the Gulf. Curr Hist 123(857):330–335. https://doi.org/10.1525/curh.2024.123.857.330
Kuwait Education Information (n.d.) List of universities in Kuwait. https://www.kuwaiteducation.info/universities
Kuwait Government Online (n.d.) Civil universities list. https://www.e.gov.kw/sites/kgoenglish/Pages/CitizensResidents/Education/Universities.aspx
Lakhno M (2024) Green or green-washed? Examining sustainability reporting in higher education. Higher Educ Quart 78(3):1099–1122. https://doi.org/10.1111/hequ.12513
Lee SS (2021) A precarious balancing act. Contemp Arab Affairs 14(1):113. https://doi.org/10.1525/caa.2021.14.1.113
Magd H, Bystrov I (2021) Commitment of higher educational institutions to sustainability agenda: comparative study of GCC and ASEAN. Global Bus Manag Res 13(3):44–56
McCowan T (2019) Impact on the SDGs. In: Higher education for and beyond the sustainable development goals. Palgrave Macmillan, pp 211–248
McCowan T (2020) The impact of universities on climate change: a theoretical framework. Transforming universities for a changing climate working paper series, No. 1. https://www.researchgate.net/publication/343889746_The_impact_of_universities_on_climate_change_a_theoretical_framework
Ministry of Education and Higher Education, State of Qatar (2024) Higher education in Qatar. https://www.edu.gov.qa/en/Content/HigherEducationinQatar
Ministry of Education, Kingdom of Saudi Arabia (n.d.) University Education. University education. https://www.moe.gov.sa/en/education/highereducation/Pages/default.aspx
Ministry of Higher Education and Scientific Research, United Arab Emirates (2023) Higher education institutions. https://www.mohesr.gov.ae/En/OpenData/Pages/home.aspx
Ministry of Higher Education, Research and Innovation, Sultanate of Oman (2022) Higher education. https://www.moheri.gov.om/InnerPage.aspx?ID=
Mosly A (2022) Education in the GCC: developments and trends. Gulf Research Center. https://www.grc.net/documents/6347b678f35acEducationintheGCCDevelopmentsandTrends.pdf

Robinson K (2022) What is the kafala system? Council on Foreign Relations. https://www.cfr.org/backgrounder/what-kafala-system

Sachs JD, Lafortune G, Fuller G (2024) The SDGs and the UN summit of the future. In: Sustainable development report 2024. Dublin University Press. https://doi.org/10.25546/108572

Shafi N (2024) Gulf floods remind us that climate mitigation cannot wait. World Economic Forum. https://www.weforum.org/stories/2024/05/floods-in-the-gulf-remind-us-that-investing-in-climate-mitigation-cannot-wait/

Stein S (2024) Universities confronting climate change: beyond sustainable development and solutionism. Higher Educ 87:165–183. https://doi.org/10.1007/s10734-023-00999-w

The Economist (2018) From pearls to black gold: how oil transformed the Gulf. Economist 427:9097

Umar S, Rahman MM (2023) International student migration to the GCC states: a comparative study. Migr Develop 12(1):1–30. https://doi.org/10.1177/21632324231194767

Umar T, Umeokafor N (2022) Exploring the GCC progress towards United Nations sustainable development goals. Int J Soc Ecol Sustain Develop 13(1):1–32. https://doi.org/10.4018/IJSESD.2022010105

UNESCO Institute for Statistics (2025) Enrolment and attendance. https://databrowser.uis.unesco.org/browser/EDUCATION/UIS-EducationOPRI

Zaidan E, Al-Saidi M, Hammad SH (2019) Sustainable development in the Arab world: is the Gulf Cooperation Council (GCC) region fit for the challenge? Develop Pract 29(5):670–681. https://doi.org/10.1080/09614524.2019.1628922

Zaidan E, Belkhiria E, Wazen C (2023) Universities of the future as catalysts for change: using the sustainable development goals to reframe sustainability—Qatar University as a case study. In: Al-Maadeed Masa, Bouras A, Al-Salem M, Younan N (eds) The sustainable university of the future: reimagining higher education and research. Springer, pp 1–23

Elizabeth Buckner is Associate Professor of Higher Education at the Ontario Institute for Studies in Education (OISE) at the University of Toronto, where she also holds the Tier 2 Canada Research Chair in Higher Education for Sustainable Global Development. Her research examines how global trends affect higher education and one strand of her work focuses on how universities are integrating sustainable development into their missions and operations.

Karma Salloum holds an Honours Bachelor of Arts degree in International Development Studies from the University of Toronto. Her research interests include governance, participatory development, education, and social development policy.

Khalaf Al'Abri is an Associate Professor of Education Systems and Policies at Sultan Qaboos University, Oman, where he also serves as Head of the Department of Educational Foundations and Administration. His research focuses on higher education policy, governance, sustainability, and internationalization in the Gulf region. He has led several national and international research projects on sustainable development, quality assurance, and global citizenship education. Dr. Al-Abri has published extensively on university rankings and higher education reform. He is President of the Gulf Comparative Education Society and collaborates with UNESCO and other international partners to promote education for sustainable development.

Part I

Adapting Higher Education for Sustainable Development: Innovative Curricular and Pedagogical Practices

Emirati Students and Climate Change Adaptation: Green Activism in the UAE?

2

Tilde Rosmer

Abstract

How are the leaders' discourse and efforts on climate change received by the Emirate public in this top-down organized society? Most research on environmental engagement focuses on European and North American countries and the small body of literature on environmentalism in the Middle East focuses on either policy or on practical possibilities and constraints. The present chapter fills both these gap by discussing the relatively lacking environmental knowledge and engagement among young Emirate students and argues that this is related to their high level of trust in their leaders, their low exposure to hands-on experiences with nature and to sustainable activities, as well as to the de-contextualized approach in the teaching of these issues. To remedy the situation, it is here suggested to produce contextualized research and teaching materials, arrange student field trips, and in general to approach the topic from the local, in this case tribal and Muslim, culture.

Keywords

United Arab Emirates · Climate change · Students · Awareness · Engagement

T. Rosmer (✉)
College of Humanities and Social Science, Zayed University, Dubai, UAE
e-mail: tilde.rosmer@zu.ac.ae

2.1 Introduction

The rulers of the United Arab Emirates (UAE) are quick to respond to the latest global trends and seem to take the reality of climate change and the future limitations of their petroleum-based economy seriously. The country has over the last decades initiated several renewable energy projects, such as two large-scale solar panel plants, and since 2010, they have hosted the *International Renewable Energy Agency*. In 2005, the UAE ratified the Kyoto Protocol to the UN Convention on Climate Change, and it has since signed and ratified the Paris Agreement. Sheikh Mohammad bin Rashid al Maktoum, Vice-President and Prime Minister of the UAE and Ruler of Dubai, has declared two ambitious goals for 2050: to have the lowest carbon footprint per capita globally (contrasted with Dubai having the highest such footprint when he took office in 2006), and to generate 75% of Dubai's power output from clean energy (Dubai Electricity and Water Authority 2020). Further indicating the country's interest in sustainability, in 2022 Dubai hosted the EXPO with a focus on sustainability and in November 2023, it hosted COP28.

Emirati society can be described as organized in a top-down manner based on a tribal patriarchal structure with un-elected sheikhs as political rulers, combined with a system of western-inspired governmental institutions and organization (Heard-Bey 2005). In this context, policies

K. Al'Abri and E. Buckner (eds.), *Higher Education and Sustainable Development in the Arab Gulf States*, Sustainable Development Goals Series, https://doi.org/10.1007/978-3-032-19559-3_2

and priorities are dictated from the top and therefore an important question is: how are the official discourse on and prescribed efforts to mitigate climate change and improve sustainability received, promoted or challenged by the Emirate public? So far, most studies that assess questions pertaining to climate change and a Green Shift in the Middle East and North-African region have focused on technical and economic possibilities of or limitations to renewable energy initiatives (Arouri et al. 2012; Sgouridis et al. 2013), as well as on policy, economic development, and international branding (Alhoweish and Orujov 2016; Luomi 2012). The present chapter fills both these gap by discussing the level of environmental knowledge and engagement among young Emirati citizens based on new data, asking why the Emirati leaders are seemingly more worried about climate change than their young citizens? In the present analysis, the causes for this observed distance in awareness between the leaders and the young citizens are found in the combination of a strong sense of trust in the leadership; a deficit in culturally and geographically contextualized approaches in the teaching of climate change and sustainability; and a lack of exposure to actual nature and to hands-on sustainability activities and projects. A new educational approach to remedy this situation should include new research from the local and regional context; developing teaching that includes local religious and cultural reasoning and logics for a just and sustainable lifestyle to make them relatable for Emirati students; and involve fieldtrips that take students into nature and to visit sustainability projects.

2.2 Environmental Humanities: Context is Key

It is generally agreed that a successful green shift is dependent on public support or at least on a non-antagonistic attitude toward the policies and changes needed. More specifically, the implementation of climate change adaptation strategies requires a combination of strong institutions, transparent decision-making systems, financial capital and natural resources, formal and informal networks that promote collective action, and human skills and knowledge (Gunel 2016, p. 306). Whereas the UAE has the resources to facilitate these changes, the young federation is lacking in terms of transparency, networks, and, most relevant for this research, local human know-how.

In Europe, New Zealand, and North America, environmentalism began with a focus on the preservation of wildlife combined with concern over industrial pollution by individuals, and from the 1960s this developed into a political movement led by non-governmental activists alongside the emergence of new green political parties (Milder 2017). In these contexts, the pressures for green policies and institutional changes can be described as emanating from pressures from bottom-up activism leading to top-down green policies. Often starting as grassroots "kitchen table" activism focusing on local concerns, such as the protests against Wyhl nuclear reactor project in Germany, these protests over time developed into transnational networks of environmental activism and this way local activism spurred national and international debates and events leading to larger initiatives and organizations (Milder 2017). In the German case, through these protests, the voice of "the ordinary people" lay the groundwork for what became the political Green Party (Milder 2017, p. 13). Also taking part in this development of a "greener democracy" were journalists, leftist students, and burgeoning NGOs (Uekötter 2019).

However, in the UAE the green change is almost entirely initiated from the leaders. Therefore, the overall aim of this paper is to investigate the extent of knowledge and engagement by individuals from the local Emirate community by focusing on students at a federal university. As in any societal research the local context is key, and culture is important to understand both the local mitigation of and the adaptation to climate change, in addition to analyzing the framing of its causes, threats, and remedies in a particular society: "Culture is no less central to understanding and implementing adaptation: the identification of risks, decisions about responses, and means of implementation are all mediated by culture" (Adger et al. 2013, p. 112). Hence the framing and understanding of the discourse about climate change, and the engagement this might cause, must be understood within its socio-cultural context.

Emirate society is unique in the make-up of its population where the locals are outnumbered one to ten by migrant workers. According to governmental sources, the "population of the UAE grew to about 10 million by the end of 2021. Out of this, the number of UAE citizens is around 1 million." (UAE 2022). The federal universities largely cater to the local Emirati population and not the larger expat population. Zayed University was established in 1998 and on its campus in Dubai it has only female students, to comply with the traditional and protective attitudes toward women still prevalent among this population (Alzeer 2018). As such, the women of this generation can be said to be living between modernity and tradition. The former they experience outside in a modern city structure and at the university where they undertake a modern education, and the latter they engage with at home and in their community by which they are considered "precious pearls" that need and deserve protection and have to adhere to a set of modesty rules (Alzeer 2019). Most female students are still generally in need of chaperones when going out and not all activities are considered appropriate for them (Reichenbach 2015).

While teaching courses on sustainability and climate change at the College of Humanities and Social Sciences from 2016, it became clear to me that most students had little knowledge about the causes and effects of climate change. Some became interested and engaged while studying, others were bored and expressed that this is "taken care of" by the leadership and institutions, and yet others expressed feeling that this was beyond their grasp and/or reach. Hardly any were engaged in any form of what would be labeled as environmental activism in Western countries. This fits with earlier observations by Ouis (2002) that local initiative and engagement in climate change and sustainability among nationals is largely absent, and a decade later Luomi also observed that there is "… no profound environmental awareness among most UAE citizens" (Luomi 2012, p. 342).

One part of the explanation for this seemingly lacking engagement is the focus on activism as per Western standards. Whereas it is largely agreed that environmental activism forced climate change onto the political order in Europe and North America, "activism" is by itself a contentious word and one that few in the UAE would use to describe their social engagement. "Activism" is per the dictionary defined as "a doctrine or practice that emphasizes direct vigorous action especially in support of or opposition to one side of a controversial issue" (Merriam-Webster 2020). In other words, actions aimed to force a change related to political, economic, social, or environmental issues. Action is by necessity based on awareness, which in turn is based on knowledge of whatever issues and/or circumstances where the change is sought. However, when asked about their environmentally related activism, local interviewees have responded negatively or reservedly, pointing out that they are not in any way engaged in any form of "activism." Similarly, in my classes on sustainability and climate change students emphasized that many of the methods and approaches of Western environmentalist activism is not an option here—public display of activism or any protest is illegal in the UAE.

This is another reminder of the contextualized approach needed that necessitates fieldwork methodology and, also, to employ a suitable terminology in discussions with the interviewees, without losing the meaning of the activities studied or the opportunity for a comparative perspective. Arguing for a contextualized approach when studying climate change and sustainability is not new (Adger et al. 2009; Wilhite 2014; Hanson-Easey et al. 2015), but still needs to be emphasized, to avoid cross-continental and cross-cultural generalizations and to inquire into the local circumstance in which attitudes and understanding is shaped and framed.

While it previously was argued that lack of information or so-called NIMBYism (Not In My Back Yard) were the reasons behind meager support for green policies, researchers have since criticized these explanations (Owens and Driffill 2008; Zaunbrecher et al. 2016), and argue that the picture is more complex: support and acceptance for different adaptation and mitigation efforts are highly dependent on values, norms, and culture and must be understood within the relevant local contexts (Adger et al. 2009). This is particularly evident in the present case study where religion and traditional patriarchal culture, mixed with the modern tool of social media, are identified as important in the shaping of environmental interest in and engagement with climate change issues among the interviewees.

2.3 Data and Methodology

The ethnographic research conducted for this paper is composed of class room observations, ten in-depth interviews, three focus groups, a survey of 138 students (conducted with Dr. Gergana Alzeer) in 2022, and participatory observations of student engagement (2016–2022), such as taking students to participate in the Emirates Environmental Group's (EEG) Inter-College Speaking Competition from 2016 to 2021. The participants were chosen using an incidental and convenient sampling method among Emirati students at the federal university. All participants were Emirati women students between the ages of 19–23 and all of them are enrolled in undergraduate programs. The students mainly live in Dubai while some participants are from other Emirates, such as Abu Dhabi, Sharjah, and Umm Al Quwain. The first focus group was conducted in December 2020, the second in April 2021, and the third in May 2023, all with seven to nine participants and all lasted 90–120 min. After the focus groups, the participants were asked to write about their experience and interviews were conducted with some of the participants, allowing for a deeper understanding of their individual experiences and understanding of the topics. All the student interviewees are anonymized unless they have an official role and are happy to be identified.

To illustrate the relevance of a localized approach, the seeming deficit in concern and knowledge about sustainability found among most students interviewed for this research is juxtaposed by the trajectories of a couple of Emirati sustainable action trailblazers whose sources of motivation for their environmental engagement confirm the appropriateness of a contextualized approach. Habiba Al-Marashi is the initiator and director of EEG that is the only Emirati environmental NGO, and the self-entitled "Green Sheikh" Sheikh Abdul Aziz al Nuaimi is an active public speaker and facilitator of activities related to environmental issues in the UAE. By focusing on local religious and cultural reasoning for a sustainable lifestyle, both the student interviewees and the trailblazers teach us how to make the topics of climate change and sustainability relatable for Emirati students.

2.4 Findings

In this section, I discuss key findings from the interviews.

2.4.1 A Knowledge Gap

From empirical data gathered in the focus groups and survey conducted for the present research, students displayed a generally low level of knowledge about climate change's causes and effects. They were able to identify global warming as related to and/or a cause of climate change, but not able to explain why or how. Moreover, they could not identify with certainty the effects climate change has had/would have on their country. In addition, students generally displayed little knowledge about local organizations and initiatives established to combat climate change or to ensure sustainable adaptation. In their responses to the question "name any environmental organizations in the UAE you know" the students responded with a wide variety and with little internal consistency, which tells us that they do not have a base-line perception of which local or national organizations or initiatives operate in this field. In the survey, of the 138 responses, the EEG received the highest number of mentions which was (only) eight; Masdar sustainable neighborhood and research center in Abu Dhabi received seven mentions; Beeha waste colleting company in Sharajah received six mentions; and Emirates Marine Environmental Group received five mentions. However, thirty-nine respondents wrote "None" and the remaining 96 did not respond to this question.

When asked where they had learnt about "climate change" and "sustainability," over 50% replied at school, less than ten percent responded from their parents, and 30% replied at university, however these students were not able to identify in which course or provide much more specific information. 30% seems low considering that in theory, all Zayed University students must study one course on sustainability in their undergraduate degree, therefore, if it is representative that only 30% remember this course, and of these only some remember some of its content, this indicates that this learning has not been absorbed satisfactorily.

In my observations of the EEG's inter-college speaking competition from 2016 to 2021, one general impression was that most of the competing teams, representing the diversity of local and international universities and colleges in the UAE as well as some from the wider Gulf region, focused on global and theoretical problems or solutions related to the environment. However, very few if any teams presented research of issues pertaining to the UAE. This seems to indicate that participating students associated climate change and sustainability with something theoretical and/or removed from their context and everyday life.

2.4.2 Sources of Motivation: Patriarchy, Religion, and Nature

In the focus groups, the sources of the environmental awareness among the students were explored in more detail and it became clear that for these students their perceptions regarding climate change and sustainability are highly intertwined with religious values, cultural practices, and tribal hierarchies. Patriarchal tribal systems of governance and Islam are the cultural elements that are intertwined in this country's history and, as such, utilized purposefully by the leadership (Ouis 2002; Heard-Bey 2005). It became evident that the political leadership plays a significant role in setting the agenda also for this young and educated generation—due to the strength of the patriarchy which allows the leaders to be trusted, admired, and venerated. To illustrate this veneration, in *all* my online classes during the COVID-19 pandemic, there were several students who used a photo of the United Arab Emirates' founding father and first President, Sheikh Zayed bin Sultan Al Nahyan, as their profile photo, as well as some using photos of other current leaders.

In the focus groups, students spoke about the UAE's founding father, Sheikh Zayed, as an inspiration for their (potential) interest in environmentalism. They recalled having learnt about how he had planted trees and distributed trees to be planted by the locals themselves, as well as how he cared for the local nature both by preserving it and by improving it. Indeed, the concept of "greening the desert" was a pillar in Zayed's modernization approach from the 1970s. Planting trees and other concrete greening is part of Emirati nationalism and, as such, constitutes part of the legitimization of power for the ruling Sheikhs and the political system of paternalism (Ouis 2002). Sheikh Zayed is portrayed as "the leader who prefers green color," and "promoted as one who loves nature and who alone devoted himself to the greening of the desert" (Ouis 2002, p. 338). He saw the actual greening of this new country as a way of changing it from a desert-dominated, traditional, and largely Bedouin society into a modern state with state-of-the art education and healthcare, and adorned with parks, avenues, and gardens with green grass and trees shading the roads. His inspiration was the modern green cities of Europe combined with his desire to improve the standard of living for his people and with his Islam-inspired love for nature (Ouis 2002). Since the 1970s, millions of trees have been planted by the government and organizations, such as in Sheikh Mohammed's recent initiative to plant one million trees before the World Expo (Case Study Middle East 2016).

Perceiving environmentalism through the actions of the founding leader is not reserved to this generation students, as evidenced in my interview with the trailblazing Emirati environmentalist Al-Marashi. She recalled establishing EEG in 1991 with programs aimed to create awareness among the public in UAE about climate change and other environmental challenges and works to protect the local environment from these challenges. As with the students at Zayed University, Al-Marashi's attributed a large part of her inspiration to the actions and examples of Sheikh Zayed (Inteview 2019).

By extension, our students also spoke about admiration for the current leadership and in particular Sheikh Mohammed bin Rashid Al Maktoum who, since he took over the rule of Dubai in 2006, has been active in changing the dependency on fossil fuels and limiting Dubai's carbon footprint. He, like other leaders, is active on social media informing his followers about his policies, actions, and thoughts including on the topic of climate change and sustainability. Many students mentioned his tweets and other social media posts about pollution or climate change adaptation initiatives as inspiring and informative. One example is when Sheikh Mohammed posted photos of camels who had died from consuming plastic in the desert left behind by campers: "This was a wakeup call for me" one focus group participant exclaimed. Hence, we can observe a connection between this social media reliant generation and their connection with their leaders, which seems to indicate a strong level of trust in their leadership as they accept the information relayed by the Sheikh as informed and authoritative.

The power of social media and its influence on youth in general has been discussed by many scholars (e.g., Omar and Dequan 2020; Gjylbegaj and Jararaa 2018). As discussed by Hofheinz (2011), this is the online generation for whom much of their social and political life takes place on internet fora. They are constantly "on" and news travels very fast in short sentences without much depth or context. Significantly, they do not necessarily, or if at all, follow traditional news media. Therefore, they do not necessarily follow daily or weekly developments around the world and are not very informed about anything that happens that is not reported or commented on within their circle of contacts or in their preferred sources. This is very evident in the social science courses we teach at the university where we generally experience that we often need to tell our students what has happened in the world in the last few days, before we can discuss these events. For climate change awareness, social media seems to have a positive effect in terms of spreading information, especially as we now know that the majority follow their leaders who post on this topic, however, it does not seem that this new knowledge in general leads to any organized manner of engagement to protect the environment.

Another source of inspiration for caring about the environment the students spoke about is religion, referring to the responsibility of mankind to take care of the creation of God. In the survey, we asked the students what their main motivation to support the environment was providing them with these options: Islam; moral, and ethical values; cultural values; feeling responsibility toward your country and the planet; and feeling responsibility toward the next generation. Most responded that their motivation was made up of a mix of religion and responsibility as a citizen to this country, which by extension included the next generation. Sheikh Zayed himself often invoked religion in his nature-focused greening policy of the Emirates in the 1970s onwards (Ouis 2002). Green is considered the color of Islam and of paradise, described as lush and bountiful, and, in the desert of the Gulf, green is the relatable color of life.

The follow-up question was "How does your strongest motivator from the question above influence you?" and many answered that Islam or religious values were the most important and that they taught them how to act. For example, one student responded: "Islam, moral and ethical values… these are my strongest motivators because it is more about a mindset since I was a child, and that what helps in the influence, because if the motivation started from a younger age, it will develop and increase while growing up." Another responded simply: "Our religion asked us to protect the earth."

When I asked Al-Marashi (2019) about what had inspired her to establish the first and only environmental NGO in the Emirates, she also spoke about the motivation from her religion combined with the inspiration from her parents stating, "It all starts at home". From when she was a little girl, she was taught not to waste anything or be greedy—for example to only serve herself with what she could eat during family meals and finish what she had on her plate. Her family was conscious about using and reusing resources, before recycling was even a term they knew about. Al-Marashi connects the attitude of avoiding wastefulness and caring for her surroundings to Islam and described being raised to respect all living creatures and natural resources that are the "bounties of Allah" and emphasized that we will be held accountable for our interaction with the world on the Day of Judgement. She smiled when she recalled being known as the one in the extended family who was against using disposable cutlery or plates from a young age. She explains that back then, her waste-avoiding approach was based on a mixture of economic and conservational ideas and not environmentalism as such. "Why spend money on new plastic or styrofoam plates, and cups when you can buy and reuse proper plates and cups?" she remembers asking her family members (Interview 2019).

Similarly, some students responded that religion taught them not to waste: "Islam—since it's our basic principle not to overspend like on food or clothes when others need it more," one student wrote while another wrote "Islam teaches us to not waste water, food, etc. I believe Islam really does cover every aspect of my life and in order to respect my religion, I have to respect the planet, as I have a responsibility to make sure that I do not impact anyone or anything negatively."

From the focus groups, it was noticeable that in general the students spontaneously talk more about pollution, waste, and recycling, rather than about climate change and energy sources. Most of their institutive examples were from things observed in real life—such as rubbish in the desert or parks, lack of recycling bins, or alternatives to plastic bags. They drew from their daily lives and observations and not from what they had been taught about climate change or heard about local initiatives to combat it. It seems clear that the theoretical approach to teaching climate change and sustainability has not been pegged on hooks that make sense to these students who therefore has not necessarily absorbed this learning as it is not (directly) related to their everyday lives, in the same way that pollution is.

The self-proclaimed "Green Sheikh," Sheikh Abdul Aziz al Nuaimi (2020) from the royal family in Ajman, explained how his engagement for the environment began when he was a young man working in the petroleum sector and there was a large oil spill on the coast of Ajman. In January 1998 a 11,000-ton oil barge ran aground and approximately 4000 tons of oil spilled on beaches and threatened marine and bird life. The leaked oil in the sea along the Emirate of Ajman posed a threat to its water supply, since the United Arab Emirates relies on desalinated seawater. Assisting in the clean-up was a turning point for Nuaimi, leading him to become an environmentalist. Observing how little awareness there was in the UAE about the environment, he began his activities and today works fulltime as the Green Sheikh, arranging sustainable trips, giving lectures, participating in social and other media discussion on sustainability, and advising the government of Ajman on environmentalism. Nuaimi mostly uses Arabic in his communication and is thereby accessible to the local community and provides information in a culturally adapted manner. Nuaimi is therefore able to connect with the parent and grandparent generation of our students, whose level of English is much less advanced, if at all present, compared with the youth. This is another significant contextualized point when it comes to increasing awareness among the local population in the Emirates—material needs to be available in Arabic (and other languages such as Hindi an Urdu to reach the expat communities).

As for the students' interaction with and experiences of nature, it is clear from the survey and the focus groups that the majority spend around or less than one hour outdoors daily. When asked about where they spend time in nature, the places they mentioned were first were the family gardens and public parks, followed by the beach and the desert, and only a few mentioned the mountains. Thus, from this research it seems that a significant proportion of the students interact with made green nature, such as planted and watered gardens and parks. Others participate in the common practice of large family gatherings in the desert on the weekend evenings where they cook traditional meals and socialize. As observed by several of the student interviewees, sadly many of these gatherings leave behind rubbish in the desert. All in all, this group of Emirati students do not interact significantly with the local nature, and only exceptionally few have been on a hike in the mountains or been on a canoe trip among the mangroves. The reasons for this are the combination of new lifestyles and habits stemming from the modern comfortable life of this generation, involving driving cars everywhere and keeping cool in the hot weather in air-conditioned spaces, in conjuncture with the views on what is considered proper and not proper for these young women in terms of activities based on the conservative local culture (Reichenbach 2015). In a campus exhibition in 2019 about the student garden club *Kulna* ("All of Us"), a project initiated by design Professor Chung-Shin (2019) with the aim of emphasizing the connections between the individual, the environment, and the society to teach and facilitate social engagement, one poster highlighted comments challenging behavioral expectations of young Emirati women directly related to interaction with nature:

2.4.3 Who Says Good Girls Don't Get Their Hands Dirty?

A common stereotype many of our female students have had to overcome is that "girls don't like to get their hands dirty or work outdoors with the soil." This story follows the experiences of Kulna's student leader Almanar. Almanar's journey has taken her from a rather quiet student who found a special place in the garden to a strong advocate for sustainable gardening practices, environmentalism, a graduate of an international permaculture design course, student club President and driver of social change. When she first began working with Kulna, Almanar recalls the reactions she would get from peers and staff, many who believed that outdoor gardening was not "girls' work." But for Almanar, working with Kulna has been hard, authentic, collaborative and meaningful and filled with self-discovery and community building, you know "girls' work."

Most of the students who were engaged in the garden club when I did my interviews were from the College of Nature and Health Sciences, but also a few from other colleges were involved. The project focused on food security and waste, and aimed to spread awareness about how to make compost and use water sustainably in this warm region when growing fruits, vegetables, and herbs. They connected their learning to health and wellbeing. The actual garden was on campus by one of the exits gates so that many students and staff passed by it on a daily basis. The engaged Emirati students emphasize the importance of learning from the past and had involved their mothers and grandmothers to teach them about local herbs and food and their usages. One of the students (interview) who was active in Kulna told me that she gradually became more environmentally aware and started making small changes to her life, such as having a refillable drinking bottle and avoiding buying new clothes as much as possible. She described how taking courses that taught environmentalism and sustainability now started to make a big impact for her, and that she began buying second-hand clothes when she was abroad, something she described as "unheard of" among the local population. Kulna serves as an example of a successful hands-on learning experience at the university wherein the students engaged not only with nature, but also integrated their culture and traditions and, as such, this project should serve as an example to follow in future practical teaching of sustainability.

2.5 Conclusions and Recommendation for New Teaching Approaches

Considering that the level of knowledge and general awareness concerning climate change and sustainability among our students is not high when compared with the engagement and efforts of the UAE leadership, this research found that the reasons for this gap stems from the combination of the local political culture and the de-contextualized and lacking nature-experience in the teaching these subjects. In the patriarchal and tribal social structures that underpin the modern institutions and city-state construction of the UAE, it is clear from this research that Emirati youth is somehow pacified in terms of acting vis-à-vis climate change and sustainability by the combination of the hierarchical decision-making system and their trust in their leaders. The patriarchal value-system does not have room for public protest and therefore all room for political involvement is prescribed, leaving little room for impulsive or grassroots engagement. Moreover, tribal traditions including respect for and trust in the perceived responsible leaders and elders that helps explaining the lack of a perceived need among students to engage vis-à-vis climate change and sustainability, especially as their leaders are already doing so. In other words, considering the impressive development of the country since the 1970s and the good life available in the country now, the majority of Emirati youth in this research expressed feeling in good hands and feeling safe under the guidance of their leaders. As such, their positive experiences of the development of the UAE thus far serves to prove that their leaders know what they are doing and that they are doing what they are doing with the best interest of their population in mind.

Based on the analysis of the empirical findings, this paper suggests that future work for higher education in the UAE aimed to improve the youth's involvement in the country's climate actions should be to develop contextualized programs that incorporate the religious Islamic aspects of caretaking for nature; the idea of paradise as a green ideal; and the tribal structure of society with the importance of the role modeling by the leaders, as evidenced in the students' social media following of their leaders. Moreover, such programs should increase students' hands-on experiences and participation with local environmental organizations, thus making the learning practical and relevant both for the students in a way that will also benefit the country practically. Thus students' feelings of ownership regarding the local nature and environment will increase, wherein they will hopefully become more engaged and have a stronger sense of agency as regards sustainability, instead of relying so heavily on the leadership to make the correct decisions and take action. The Emirati population needs to be more involved in sustainable actions as ultimately, in this place with such a high percentage of an expatriate transient population, the Emirati nationals are here to stay and have both ownership over this place as citizens and landowners. Today's youth have a future to invest in the UAE and federal universities need to prepare them for this complicated and complex task based on who they are and where they are, by producing localized research and new contextualized teaching approaches.

References

Adger NW, Lorenzoni I, O'Brien K (2009) Adaptation now. In: Adger NW, Lorenzoni I, O'Brien K (eds) Adapting to climate change: tresholds, values, governance. Cambridge University Press, Cambridge, pp 1–23

Adger W, Barnett J, Brown K et al (2013) Cultural dimensions of climate change impacts and adaptation. Nat Clim Change 3.112–117. https://doi.org/10.1038/nclimate1666

Alhoweish B, Orujov C (2016) Promoting an effective energy efficiency programme in Saudi Arabia: challenges and opportunities. In: Raouf MA, Luomi M (eds) The green economy in the gulf. Routledge, New York, pp 100–123

Almarashi H (2019) Interview with author. Emirates Environmental Group's office, Dubai

Alzeer G (2018) Women's spatial experiences in Emirati higher education: between modernity and traditions. Br J Sociol Educ 39(8):1175–1194. https://doi.org/10.1080/01425692.2018.1478278

Alzeer G (2019) Spatializing higher education: Emirati women learners' hot and cold spaces. Int Stud Sociol Educ 28(1):46–62. https://doi.org/10.1080/09620214.2018.1470021

Arouri MEH, Youssef AB, M'henni H, Rault C (2012) Energy consumption, economic growth and CO_2 emissions in Middle East and North African countries. Energy Policy 45:342–349

Case Study Middle East (2016) One million trees: a 2020 vision for the United Arab Emirates. Land Life Company Magazine1. http://magazine.landlifecompany.com/one-million-trees#:~:text=The%20UAE%20has%20a%20long,World%20Expo%202020%20in%20Duba. Accessed 15 Jan 2021

Chung-Shin Y (2019) Interview with author. Zayed University, Dubai
DEWA (Dubai Electricity and Water Authority) (2020, 5 January). Renewable energy shaping the future of sustainability. https://www.dewa.gov.ae/en/about-us/media-publications/latest-news/2020/01/renewable energy-shaping-the-future-of-sustainability
Gjylbegaj V, Jararaa O (2018) The effects of social media on youth's interpersonal communication in UAE. IJASOS Int E-J Adv Soc Sci 4(10):23–29
Günel G (2016) The infinity of water: climate change adaptation in the Arabian Peninsula the infinity of water. Public Cult 28(2):291–315
Hanson-Easey S, Williams S, Hansen A, Fogarty K, Bi P (2015) Speaking of climate change, a discursive analysis of lay understandings. Sci Commun 37(2):217–239
Heard-Bey F (2005) From trucial states to United Arab Emirates. Int J Commun 5:1417–1434
Hofheinz A (2011) Nextopia? Beyond revolution 2.0. Int J Commun 5:1417–1434. http://ijoc.org/index.php/ijoc/article/view/1186
Luomi M (2012) The gulf monarchies and climate change: Abu Dhabi and Qatar in an era of Natural Unsustainability, Hurst
Milder S (2017) Greening democracy: the anti-nuclear movement and political environmentalism in west Germany and Beyond, 1968–1983. Cambridge University Press, Cambridge
Nuaimi AA (2020) Book talk, Emirates airline festival of literature, Dubai
Omar B, Dequan W (2020) Watch, share or create: the influence of personality traitsand user motivation on TikTok mobile video usage. Int J Interact Mobile Technol 14(4):121–137. https://doi.org/10.3991/ijim.v14i04.12429
Ouis P (2002) Greening the Emirates the modern construction of nature in the United Arab Emirates. Cult Geogr 9:334–347
Owens S, Driffill L (2008) How to change attitudes and behaviours in the context of energy. Energy Policy 36:4412–4418
Reichenbach A (2015) Gazes that matter: young Emirati women's spatial practices in Dubai. Urban Anthropol Stud Cult Syst World Econ Develop 44(1/2):113–195
Sgouridis S, Griffiths S, Kennedy S, Khalid A, Zurita N (2013) A sustainable energy transition strategy for the United Arab Emirates: evaluation of options using an integrated energy model. Energy Strat Rev 2(1):8–18
UAE (2022) Population and demographic mix, the United Arab Emirates' governmental portal. https://u.ae/en/information-and-services/social-affairs/preserving-the-emirati-national-identity/population-and-demogr aphic-mix. Accessed 1 April 2024
Uekötter F (2019) Stephen milder: greening democracy: the anti-nuclear movement and political environmentalism in west Germany and beyond, 1968–1983. Am Hist Revi 124(1):1452
Wilhite H (2014) The problem of habits for a sustainable transformation. In: Syse KVL, Mueller ML (eds) Sustainable consumption and the good life: interdisciplinary perspectives. Routledge, London
Zaunbrecher BS, Bexten T, Wirsum M, Ziefle M (2016) What is stored, why, and how? Mental models, knowledge, and public acceptance of hydrogen storage. Energy Proced 99:108–119

Tilde Rosmer is an Assistant Professor at the College of Humanities and Social Sciences at Zayed University, Dubai. Her research has focused on collective identity formation and religio-political and cultural movements by and for Israeli Jews from the Middle East and North Africa, and on Palestinian citizens of Israel. In 2022, she published her monograph The Islamic Movement in Israel with the University of Texas Press. Currently, Dr Rosmer is conducting research in the emerging field of Environmental Humanities, investigating the levels of awareness and knowledge of sustainability and climate change among Emirati youth. She is also part of a research group that investigates and has published on Diversity, Equity and Inclusion in the context of the UAE.

Universities as Living Laboratories in the Gulf Region

3

Anna Grichting Solder

Abstract

Addressing the complex and interconnected challenges of our territories and societies in the future requires new approaches to education that contribute to more resilient environments and communities, fostering creative, agile, and interdisciplinary thinking. This implies breaking down the silos that separate our educative disciplines as well as reaching out to society and government institutions. The University campus has been referred to as a microcosm of the broader community and is positioned to foster new social and cultural values that can transform and regenerate our declining ecosystems. This paper examines how living laboratories in Gulf region universities can support education and research on sustainable development and climate change. It builds on research and teaching undertaken at Qatar University, which resulted in the creation of a living laboratory as part of an international research project focused on the food-water-energy nexus. It also questions whether living laboratories can help universities to adapt more swiftly to the needs of society.

A. Grichting Solder (✉)
University of Vermont, Burlington, VT, USA
e-mail: anna@annagrichting.com

Keywords

Living laboratories · Nexus approach · Interdisciplinarity · Regeneration · Nature-based solution · Food-water-energy nexus

3.1 Introduction and Background

Addressing the future complex and interconnected challenges of our territories and societies requires new approaches to education that can contribute to shaping more resilient environments and communities, fostering creative, agile, and interdisciplinary thinking. Humans need to be educated to anticipate and respond to rapidly changing systems and societies and new situations such as pandemics and climate change, which we have seen deepen divides and social injustices. Such an effort implies breaking down the silos that separate our educative disciplines as well as reaching out to society and government institutions. It is widely known that universities are key players in terms of the implementation and further development of sustainability, with some having the potential to act as "living labs" in this rapidly growing field (Filho et al. 2020). Cities have a long history of being seen as laboratories or sites of social experimentation. Urban planners, activists, and other policymakers are increasingly treating cities as "living laboratories" of sustain-

K. Al'Abri and E. Buckner (eds.), *Higher Education and Sustainable Development in the Arab Gulf States*, Sustainable Development Goals Series, https://doi.org/10.1007/978-3-032-19559-3_3

ability and urban laboratories are emerging in numerous cities around the world as a strategy for local governments to partner with public and private property owners to reduce carbon emissions, while simultaneously stimulating economic growth (Evans and Karvonen 2013).

According to the European Network of Living Labs, living labs are open innovation ecosystems in real-life environments using iterative feedback processes, focusing on co-creation, rapid prototyping and testing, scaling-up innovations and businesses, and operating as intermediaries among citizens, research organizations, companies, and government agencies (Lupp et al. 2021). The concept of "living laboratories" was proposed in the early 1990s to describe students' experimentation in a Philadelphia neighborhood (Bajgier et al. 1991), and it was further developed by Mitchell (2003) at the MIT Media Lab and School of Architecture in 1995 to define innovative research aimed at developing and testing information and communications technology in homes, neighborhoods, and cities. In 2005, the concept was adopted by the European Union (EU) as a progressive, experimental, and inclusive mode of planning, project design, and implementation fostering innovation in real-life and experimental settings. Today, living labs are being used to research, design, and implement nature-based solutions (NBS) to develop more collaborative and sustainable solutions for reducing exposure to natural hazards and vulnerability to events, such as increased flooding in a changing climate (Lupp et al. 2021).

Several projects for nature-based solutions funded under the European Horizon 2020 use participatory processes and the living lab for the selection, co-design, implementation, monitoring, and evaluation of their nature-based solutions. Citizen participation methodologies for the design and implementation of solutions, with collaboration among different public and private actors and citizens, are recognized as an efficient tool in environmental governance (Bodin 2017). Deepening the agenda brought by participatory approaches and co-creation by citizens in the processes around selecting, designing, and implementing nature-based solutions needs to be extended to encompass "*co-nature'ing*". Co-nature'ing implies linking to, and associating with, nature. While it draws on awareness, it also reflects a choice of caring for—and recognizing—being part of our wider ecosystem, where nature's resources such as air, water, soil, flora and fauna, are not taken for granted (Global Forum 2022).

The university campus has been referred to as a microcosm of the greater community (König 2013), which puts it in a key position to foster and encourage new social and cultural values that are backed by students, who will be the ones to significantly alter their society. A living laboratory can be a way to reach out across the university communities, creating partnerships, and offering unmatched opportunities to teach, do research, demonstrate, and learn about all facets of sustainability and environmental challenges, as well as experimenting with new forms of governance and implementation, including bottom's up and top-down approaches.

Universities play a clear role in spreading information and developing highly qualified employees to meet economic, environmental, and societal needs, and they are crucial institutions in processes of social change and development (Brennan et al. 2004). Universities can discover, imagine, develop, and experiment with social and technical solutions to urgent global challenges such as climate change. University living labs are defined as a social and material infrastructure within a university, enabling a multitude of place-based, interdisciplinary, and impact-oriented research, education, operations, and enterprising projects of the university in collaboration with its societal partners (Hadfield et al. 2023).

3.2 Universities as Living Labs in the GCC

This paper discusses living labs in higher education institutions of the Gulf region, as a means to implement education and research for sustainable development and climate change. There are very few examples of university living labs in the Gulf region, however, universities across the

Arabian Peninsula are increasingly implementing sustainability initiatives in the design and operations of their campuses. In research on university campuses as sustainability "exemplars" in the Arabian Peninsula (Koch 2017)three of the region's leading projects were identified as being NYU-Abu Dhabi in the Emirates, Qatar Foundation's Education City, and Saudi Arabia's King Abdullah University of Science and Technology (KAUST). The research concludes that while substantial shifts toward sustainable development in the region are underway, they have largely stemmed from market forces rather than a new environmental consciousness promoted by these three iconic universities. "Greenwashing" is not confined to corporate marketing, and actors from the non-profit sector, government, the arts, and beyond regularly use the language of sustainability to craft a positive image of their particular political or social agenda (Koch 2017).

While the university strategies in the region are discussed as a form of greenwashing—having more of a PR effect than real impact (Koch 2017)—we propose to take another tack and view these initiatives and experiments as important incubators for new models of universities that are deeply intertwined and connected to their cities and communities, collaborating with the governance institutions, the private sector, industry, NGOs, civil society, and the environment. There are a lot of young Universities in the region, and the top-down systems of governance can often make changes happen more quickly than in long-established Western Universities, so there is potential for a real development of more sustainable University campuses in the region. Newer Universities may also be more flexible and agile to respond to bottom-up initiatives to affect change. The GCC countries are generally known for top-down strategies and government-led development models but effective partnerships with a common outlook are vital to ensuring coherent, economically viable, socially equitable, and environmentally acceptable policies, plans, and programs.

Stakeholder engagement is a key component for the implementation of the living lab concept and to achieving sustainable practices. SDG17 refers to multi-stakeholder partnerships, global partnerships, and both domestic and transnational collaborations toward the effective implementation of a sustainability agenda (UN Sustainable Development Goals, 2023). The 2030 Agenda is based on an interconnected economy, society and environment, and partnering is an essential approach to SDG impact (Stibbe and Prescott 2020). The mission of higher education institutions is to positively impact wider society. While urban (or rural) living labs—as defined by the European Network of Living Labs—seek to create value that accrues primarily to society rather than to private individuals, a University tends to measure the knowledge-transfer and income it generates for the University, as well as the funding it attracts, rather than the communities it impacts (European Network of Living Labs 2024).

We will discuss two living laboratories in the region, one with a focus on technology and science (King Abdulla University of Science and Technology (KAUST) in Saudi Arabia), the other focused on food-water-energy-waste nexus and biodiversity (Qatar University QU). The cases also differ in that the KAUST living lab is a top-down implementation and vision, and part of the University's founding identity while the Qatar University living lab is a bottom up and incremental vision, developed by the author, and nourished by collaborations with authorities (public works, planning, environmental), NGOs, the private sector, civil society, and within the University, with researchers, University services, faculty, students and the entire community of workers. International research collaborations are vital, and both KAUST and QU have a strong research agenda and funding, and they also have access to well-endowed National Research Funds. Qatar University is ranked in second place (92.8) in the Times Higher Education (THE) Arabic University rankings, while KAUST is close behind in third place (92.6) (Times Higher Education 2024). Both ranked top marks with 100 points for research, while there is a large gap in the performance indicator for impact on society (knowledge transfer and impact) with QU scoring 93.5 and KAUST only scoring 50. If we look at the

SDG impact rankings, KAUST scores considerably higher (95.9) than Qatar University (75.5) in the SDG 17 on partnerships.

3.2.1 The KAUST Living Laboratory, Saudi Arabia

Probably one of the leading examples of a living laboratory in the region is the KAUST living lab. The campus presents itself as a city without boundaries on experimentation, at the intersection of land and sea, with a culture that is expanding the frontiers of innovation. (KAUST 2023). KAUST is a graduate-level international research University located on the shores of the Red Sea. The University is built on environmental stewardship principles with sustainability as a pivotal value and with a focus on six core pillars—water, energy, environment, food, health, and digital—with the ambition to become a global model of research on the world's most pressing challenges. The University educates students to become agents of change for global sustainable development. As a living lab, KAUST conducts research, funds, tests, and implements in-house sustainable technologies while welcoming external collaborators and other stakeholders to participate in this rich ecosystem.

KAUST is a private university, and it ranks 7th among the world's richest Universities, behind Harvard, Yale, Stanford, and Princeton. (Javatpoint 2024). The University opened in 2009 in the King Abdullah Economic City, a greenfield city 80 km north of Jeddah, and it is Saudi Arabia's first mixed-gender University. The campus is certified LEED Platinum and has a compact design, with waste and water recycling, siting to maximize airflow, electric bus transportation, Segways for security guards, and the use of solar power (Koch 2017). The University campus is used as a living laboratory as a testbed for research and innovation in sustainable and smart city concepts in real-life, in such fields as mobility, retail, citizen and government engagement, building management, security, and logistics. The projects are developed through design thinking workshops and journey mapping and then implemented as pilot projects. Partnerships and engagement with key industry players, technology vendors, and other Universities contribute to the co-development and co-design of concepts and ideas.

The KAUST Smart Initiative encourages residents, workers, and visitors to imagine, play, explore, experiment, and create together in an immersive, user-centric experience that applies and integrates technologies to deliver easy, intuitive experiences. Another initiative is the smart house, a joint venture between KAUST smart, facilities management, community life, and several other teams, to design and construct a liveable smart home with advanced domestic solutions that enhance home living. These include renewable energy integration, wastewater reuse, accessible living spaces, digital integration, drinking water production, and integration with other experiences such as delivery using drones and autonomous vehicles. Smart house also collects data and tests the operational feasibility of different pilot technologies.

With its location in Jeddah on the shores of the Red Sea, KAUST also applies the living laboratory concept to its "backyard sea" creating an open lab for novel approaches to coral preservation and restoration. The University created the world's first coral probiotic village in the Red Sea in 2021 as a permanent natural laboratory designed for coral reef research to conduct long-term surveys to improve knowledge on different aspects of reef ecology, diversity, and dynamics; to pilot evaluation of innovative solutions that accelerate the adaptation of the coral halobiont to stress and, consequently, to increase coral reef resilience and tolerance to impacts.

The University is highly engaged on the fostering and preservation of biodiversity—positioning the campus as a living laboratory for biodiversity and highlighting its unique location on the shores of the Red Sea. Through biodiversity research and community engagement, the lab produces cutting-edge academic research on the local environment and becoming a living laboratory for "environmental well-being". KAUST's environmental stewardship policy is based on collaboration where stakeholders, wider communities and the University develop operational policies, guidelines, and monitoring systems to manage local biodiversity through its health, safety, and environment (HSE) department. Environmental permitting, monitoring, biodiversity protection and conservation, and nature appreciation and engagement programs, working with local communities, has resulted in an increase in the size of the mangrove forests, a thriving bird population, and a keen and involved community (KAUST 2024).

At this stage of the research, little has been discovered on the evaluation and impact of the KAUST Living Lab, excepting the University ranking information. Additional research is necessary in the evaluation criteria for University Living Labs, as well as in KAUST's methods of impact assessment. The University website indicates that Saudi Affairs provides monitoring and evaluation and longitudinal assessments of the University's internal and external activities, measuring impact at programmatic, institutional, and national levels, and it confirms that KAUST's impact in Saudi Arabia is significant, and growing rapidly (KAUST 2023).

3.2.2 Qatar University Living Laboratory for the Food-Water-Energy Nexus

The Living Laboratory at Qatar University differs from the KAUST model in that it was created as a bottom-up initiative led by a faculty member and built up over several years through courses, research projects and outreach events, culminating in an international research grant with a consortium of researchers and international Universities. It also differs in that Qatar University is a public University and does not benefit from the same private endowments as the KAUST model. It is important to show the potential for such bottom-up initiatives, but also to highlight their fragility in the long term, when there is no stability of faculty tenure or research funding which can lead to a lack of continuity of the project. Qatar University campus in Doha covers approx. 800 Hectares and is Qatar's only public University, established in 1973. It has a University Farm for research located at a distance from the campus, in the rural areas. The University serves on behalf of the government and private industry to conduct regional research, particularly in areas of the environment and energy technologies.

The Qatar University Living Laboratory for the food-water-energy nexus (QU Living Lab) was implemented through the Department of Architecture and Urban Planning in the College of Engineering, in collaboration with the Environmental Science Centre, the Building and Landscaping Services, the University Communities (faculty, researchers, students, staff, workers, non-human communities) and through partnerships with the private and public sector. The project was conceived over multiple years, through different courses and research projects, and at different scales, from micro-algae research to the University master plan, also extending to the whole GCC region through comparative case studies (Grichting 2017). Combined with a multiscale approach, it is also important to involve multiple stakeholders. By bringing together international experts with local academics, researchers, students, architects, landscape architects, planners, agricultural businesses, local NGOs, local administrations, and government institutions, the living lab provided an environment to develop creative responses to the questions of food, water, energy, and waste from a systems perspective that also involves University communities and the larger society (see Fig. 3.1). This multi-pronged approach for teaching, research, and knowledge building and international experts and stakeholders.

As a partner in an international research consortium hosted by the Belmont Forum and funded by Qatar Foundation, "The Moveable Nexus: design-led urban food, energy, and water management innovation in new boundary conditions of change" (M-NEX 2024) was implemented through the Living Laboratory at Qatar University and officially launched during the international workshop hosted by Qatar University (M-NEX Doha 2019a; b). The QU Living Lab built on existing campus design projects developed with students since 2011—focused on food urbanism, blue urbanism, biodiversity,

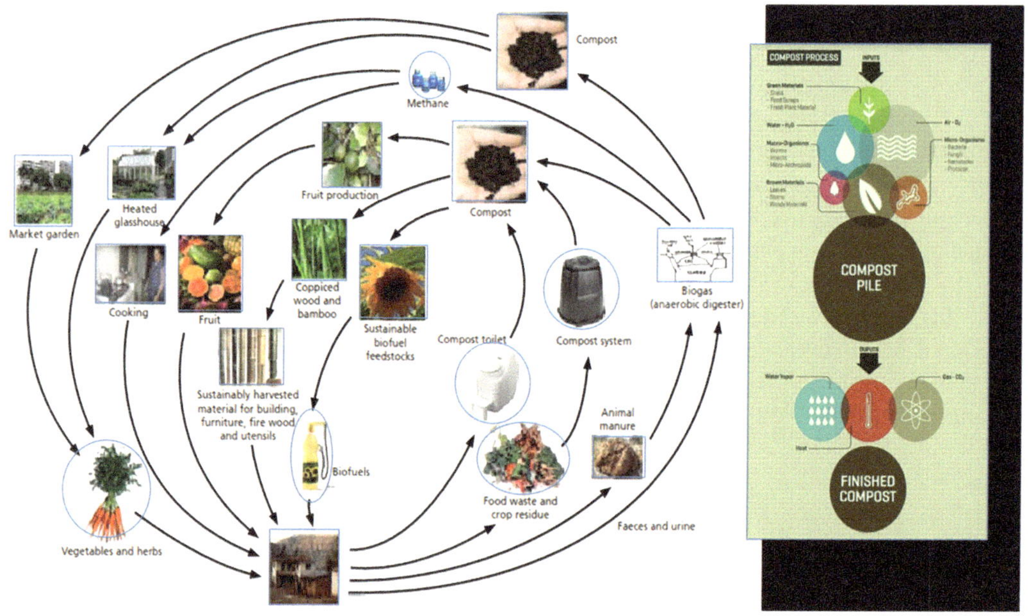

Fig. 3.1 Qatar university living lab. Thinking in systems. Composting organic waste for food, energy, and materials production. *Source* Anna Grichting

and nature-based design—as well as on graduate and undergraduate courses in urbanism and design that were developed with an integrative and systems approach to design and planning, extending into research projects, international research grants, exhibitions, public events, and participation in local, regional and international conferences. The non-academic events were key to building up the stakeholder's participation in the living lab.

The incentive to incorporate the topic of urban agriculture into the curriculum and research at Qatar University, as well as to develop a pedagogic methodology with a system thinking approach, was motivated by Qatar's reliance on imports for its food supplies, a lack of awareness of urban agriculture, and the scarcity of research and projects on growing food locally (Grichting 2022a, b). Beginning in 2011, this work was well established—if not fully understood—during the Blockade of Qatar by its neighbors (Saudi Arabia, U.A.E, and Bahrain) in 2017 which accelerated initiatives for local food production, propelling the topic of urban agriculture and vertical farming to the top of the political and research agenda. With the sudden blockade, the country was forced to develop alternative supply routes, scale up domestic production, and expand its Hamad Sea Port. Qatar become more self-sufficient, diversifying its economy and becoming largely self-sufficient for milk and dairy products, which had previously been imported from Saudi Arabia (Brookings 2023). The blockade helped to raise awareness among students on the importance of integrating food production into design projects and master plans, as well as to nexus thinking on food, water, energy, and waste.

Considered as a microcosm of the city (Konig 2013) the Living Lab at Qatar University aims to involve all the University communities including researchers, faculty, administration staff, students, technical staff, and migrant workers, non-human residents, bringing in the stakeholders from the public and private sector. The Facilities and Services Department and Capital Projects Department at QU assist with campus data, maps, and operational information during the course of the project. The Arab Engineering Bureau, the oldest and largest Qatari architectural firm was invited as an industry partner to bring its professional exper-

Fig. 3.2 Visiting Torba farms with local and international researchers. Photo: Anna Grichting

tise for the design charrettes and the design of the master plans and FEW systems. Global Farms and Turba Farm were key partners in providing performance data sources as well as their expertise in Qatar with nature-based permaculture food production and high-technology farming to develop the living lab scenarios on the QU Campus. (M-NEX 2019a, b).

3.2.2.1 The University and the Farm

One of the aims of the research project was to combine nature-based solutions and technological systems to food-water-energy nexus design to increase food production on the campus, minimizing the use of water and energy while increasing biodiversity and the quality of soil (M-NEX 2019a, b). Torba Farms is approximately the same size as Qatar University campus, and the vision is to transform the University campus grounds into a productive landscape, with food forests, permaculture gardens, biodiversity hotspots, and productive fields, using the methods and knowledge developed at Torba Farms. The technological know-how with Global Farms would be integrated into projects for productive rooftops using produced water from air conditioning, as well as for vertical farming and greenhouses on the campus.

Torba Farms is a 60-acre farm located in the Abu Thaylah region, located in the heart of one of Qatar's most well-known farming regions, famous for its sweet waters. Established over 30 years ago, it is run by Mohammed and Fatma Al-Khater, who have transformed it into one of Qatar's first organic permaculture farms. Based on organic farming and animal husbandry production principles, it combines the best of ancient and modern techniques to find viable, cost-effective, replicable, and environmentally friendly solutions that nurture both the land and its inhabitants. (Torba Farms 2023). Aligned with the M-NEX Living Lab objectives, Torba Farms strives to find creative solutions to challenging agricultural issues (food-water-energy nexus) to address issues particular and unique to Qatar's climate and terrain. The permaculture of "synergy" is at the heart of this dynamic model of farming, and it is a nature-based system that promotes biodiversity and resilient ecosystems (see Fig. 3.2).

Fig. 3.3 Qatar university living lab. Vision for green roofs. *Source* Anna Grichting

Global Farms, on the other hand, is implementing technology to increase production in the extreme Qatari climate. Global Farms is based on the concept of agriculture being more than just production, it started running many activities such as open field cultivation, greenhouse cultivation and construction, farm products marketing and so many more, and for that, it became one of Qatar‘s top technical farms in hydroponic technology. (M-NEX 2019a, b). Global Farms is working on developing sustainable agriculture processes based on complete independence from any external imports, to reach 100% Qatari production. The farm is currently cultivating more than 60 kinds of vegetables, fruits, and flowers and recycles 100 Tons of organic waste, turning it into organic fertilizer.

3.2.2.2 A Master Plan for the Food-Water-Energy Nexus

The Living Lab at Qatar University focused on developing a campus master plan that addresses the quality of life, through an increase in the quality of food, the quality of air, through reduced CO_2 and carbon capture, and increased biocultural diversity in the plant and animal species and in the communities that are involved in the production. The stakeholders and University communities are an important part of the elaboration and implementation of the project, based on the premise that it is not only technologies but also the active involvement of communities and change in behaviors and lifestyle that will build a more resilient society and sustainably affect the food-water-energy nexus, both quantitatively and qualitatively.

The QU Living Laboratory places an important focus on the soil, and how the desert soil can become fertile with efficient composting and recycling systems and efficient watering and nutrients. With that, it looks at different forms of land use, including the use of roofs as an additional layer of land use, as well as complementary systems of food production using natural (permaculture and traditional) and technological food systems—that is soil based and water-based vertical farming and green roofs (see Fig. 3.3). The project also integrates water-based urbanism to optimise water efficiency and recycling as well as nature-based solutions (Grichting, 2020).

The living lab combines technological FEW (food-water-energy) systems and interventions with natural FEW systems (permaculture) to create a resilient campus and to foster biodiversity of fauna and flora on the campus. It proposes to implement this through a plan which integrates new nexus-based architecture buildings and retrofitting existing structures. Through the modeling of the master plan the different scenarios elaborated can be quantified and qualified as to how they best achieve the objectives of the research. The design charrettes held with international researchers and local stakeholders during the International and local workshops (Grichting 2019) building on previous projects on the edible campus, resulted in design scenarios that include:

- An Edible Permaculture Garden located at the College of Engineering
- An Urban Edible Boulevard—with technology-based agriculture (hydroponic)—connecting the metro station to the campus leading to the student housing
- An Urban Edible Boulevard—using permaculture methods and soil-based—connecting the food forest and biodiversity wadi to the campus buildings
- A FEW cube building at the intersection of the two urban boulevards which would house the new agricultural college and FEW research center
- A water machine tower for the algae systems as a food-water-energy microorganism.
- The restoration of the historic campus building to house new FEW based activities
- The building of a biochar facility to produce good soil on the campus
- The creation of a permaculture food forest in the lower wadi area
- The transformations of impermeable surfaces into permeable surfaces and fertile soils
- The transformation of decorative landscapes into edible landscapes
- The use of roofs for water-energy-food production
- Increased walkability and soft mobility–electric trams, electric personal vehicles, shaded connections—to connect the FEW landscapes to the buildings and green roofs.

3.2.2.3 The QU Living Lab—From Participative Design to Policy

The Qatar University Living Lab follows the basic concept of the Moveable Nexus (M-Nex) research project based on mobilizing social and natural resources, moving stakeholders to action, and bridging local and global knowledge (M-NEX 2024). The design workshops to elaborate the Campus Master Plan for the FEW Nexus at Qatar University were accompanied by research and reflections on the policy implications and implementations to achieve sustainable practices. Building on research developed with master's students in the urban planning legislation course, which studied policies and guidelines related to implementing sustainability in the food, water and energy nexus at the global (SDGs, UN Habitat guidelines) regional, national, and University scale, we questioned how the design and urban interventions proposed for the Qatar University Living Lab could be implemented at the policy level. Bringing in stakeholders from the public sector, private sector, research, and academia on a practical project can jump-start or accelerate the discussion and communication regarding novel approaches to FEW Nexus planning. (Fig. 3.4). Visualizations of the master plan, flow diagrams and data modeling contribute to enhancing and forwarding the discussion toward policy changes in the use of different types of water, incentives for on-site treatment and waste processing, the value of local species and biodiversity, etc. The presence of professional farmers with the decision makers and engineers from public works can produce a policy that is sound and implementable.

The research also sought to develop methods of participation in the design and tools that could be used to create design processes that can include all communities and stakeholders. The proposal to use phone applications for stakeholder engagement, alongside more traditional meetings, can assist in creating an inclusive process that can both inform communities and gather knowledge and data from them, overcoming cultural and communication barriers. The community gardens proposed on the campus—designed for chil-

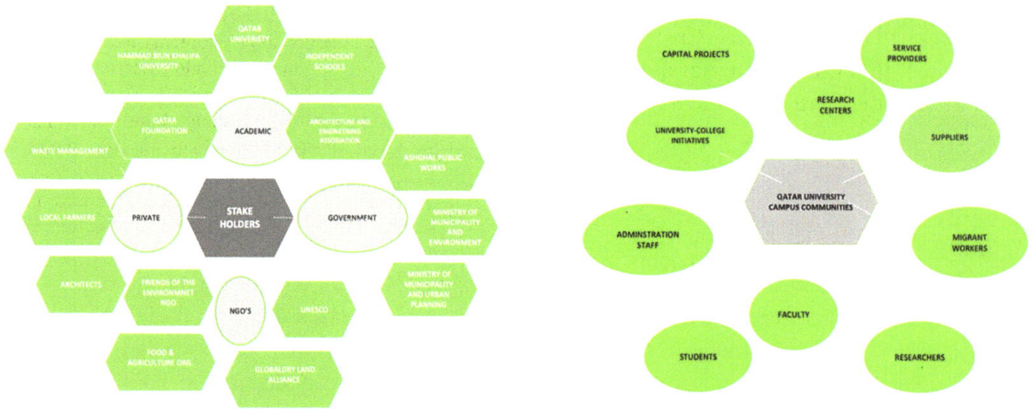

Fig. 3.4 Qatar university living lab. Diagram of stakeholders. *Source* Anna Grichting

dren, for Nepali workers, for Filipino maids, for students, for faculty for researchers—are also spaces of engagement and exchange—and the types of crops grown by different communities, the biodiversity that is generated, the social fabric that is created, all these can be measured and transformed into scientific data. From this, policies can be developed with the communities that can contribute to creating productive landscapes and community gardens that improve the resilience of the FEW nexus in the city of Doha. The model of how Qatar University engages its community in the FEW nexus design can be a model for the city and the country on how to engage the larger community. The international scope of the project, international workshops and meetings, can generate and produce policy at a higher level—international policy—and the recommendations of an international team can assist in pushing forward policy recommendations at the national level. The policy recommendations and approaches developed and discussed by the M-NEX researchers are summarized in the Table 3.1.

A thorough assessment of the current policies in Qatar regarding food, water, and energy, urban farming, and sustainable development would be necessary to identify the gaps and to identify where policies need to be modified or created.

The region's experience with one-sided development policies has shown that one cannot isolate food security from groundwater considerations, desalination from its increasing energy footprint, urban planning from air pollution and warming impacts, or coastal development from its effects on marine life and the tourism sector. Policymakers in the region are increasingly aware of these interlinkages and of the need for specific national policies and regional cooperation frameworks and a change in the mindset and culture is necessary to move towards an integrated participatory approach to policymaking geared towards sustainability and inclusion (AFED 2016). Living labs can be one way to implement participatory policymaking, especially those that are related to Universities, who have the relevant and necessary disciplines and research to bridge gaps in policymaking.

3.3 Impact, Implementation, and Evaluation of University Living Labs

Where is the resistance to implementing Living Laboratories in Institutions of Higher education? These types of projects do involve risk-taking, networking and management skills, long-term visions and incremental strategies, and interdisciplinary thinking. Universities in the region still tend to evaluate faculty on traditional methods and achievements, with a focus on specialization and peer-reviewed journals rather than rewarding faculty for addressing future environmental and

Table 3.1 Policy recommendations for the food-water-energy nexus

Theme	Problem	Approach
Water A framework for diverse water sources and integrated water systems. (the urban water machine)	• Lack of fresh water and high environmental costs of desalination • Necessity to use multiple sources of water to produce food: treated sewage effluent (TSE), salt water, produced water from air conditioners, gray water • Opportunity to build resilience and increase water security for Qatar, while minimizing environmental damage	• Downscaling infrastructure and creating decentralized water treatment plants integrated into buildings and neighborhoods • Addressing policy on TSE to allow the use in food production (not only animal feed) • Addressing cultural barriers against eating food produced with TSE by introducing labeling systems (as in organic food labeling)
Energy Incentives for renewable energy alternatives and more efficient energy systems	• Non-renewable energy is abundant and cheap in Qatar thus there is little incentive to produce renewable energies • The environmental costs—CO_2 emissions and poor air quality—must be addressed to create healthy environments and combat climate change	• Downscaling energy infrastructure and creating on-site renewable energy production systems—solar, biogas, microalgae, etc. • Reducing energy consumption in buildings, landscape, and infrastructure • Create policy incentives to implement alternative energy sources by taxing non-renewable fuels • Develop carbon capture infrastructures
Food Developing urban farming for better resource use and for productive ecosystems	• The embargo on Qatar (2017) demonstrated the fragility of relying only on external food sources • Local Qatari farms were not efficient or productive and urban agriculture was not taken seriously • A lot of land in the city is vacant, and there is a great potential for the use of rooftops • There was no practice of composting, which allows the recycling of organic waste, and the creation of fertile compost for crops • The question of food production was not linked to the question of ater, energy, and waste	• Necessity of land use planning and master planning that integrates urban agriculture in Qatar • Create incentives for rooftop greening reusing gray water and water produced by air conditioners • Implement recycling of food and organic waste to create compost for soil based agriculture • New farming technologies—aquaponics, hydroponics—associated with nature-based and soil-based food growing systems (permaculture approach) to create resilient food systems • Cultivate new types of crops, superfoods (Algae), traditional crops and species high in nutrients and protein • Adapt animal fodder to high quality and local species that increase the dairy yield and quality of the milk, reducing energy and water to produce the same amount of milk
Health Health issues due to poor diets for rich and poor Lack of knowledge on the interrelation between soil biome, gut biome and brain biome, and results on health and immunity	• Health issues due to poor diet for both wealthy communities and migrant workers • There is obesity and diabetes, due to high content in fats, free sugars and salt and illnesses due to over-consumption of animal proteins • Alternatively, there are populations such as the migrant construction workers, who suffer from high blood pressure, blood sugar issues and dehydration	• Create policies and practices that include diet and awareness on nutrition at all levels of education and in all communities • Provide access to nutritious and fresh foods to all residents through community gardens and greenhouses • Improve soil quality to increase food quality and gut health • Use public art and actions to communicate on the soil-food-gut health
Communities participative engagement of stakeholders to responsible action and sustainable solution recognizing the importance of diversity in human and non-human species	• Loss of biodiversity through urbanization, salination of ground water and sea water, desertification, etc. • Social and cultural segregation and high-income inequalities in a future knowledge economy • Social issues include the imbalance of native Qataris to foreigners, who make up 88% of the population	• Stakeholder participation as a process for designing and implementing resilient food water and energy systems for biological and cultural diversity • Bio-cultural diversity refers to the continuing co-evolution and adaptation between biological and cultural diversities and reflects people's ways of living with nature, generating local ecological knowledge and practices across generations that allow societies across the world to manage their resources sustainably while also maintaining cultural identity and social structures (Ramsar 2023) • Build on the knowledge of migrant communities on food growing and increase their knowledge on food-water-energy through public engagement and community gardens

societal challenges with agile and transformative solutions and larger societal involvement. Appraisal is generally more about the publication impact factor than the real impact on society, community, and environment, and we would need more incentives to encourage faculty to engage in such large and long-term visions. We also need more research to open up the discussion on these topics and other questions. Can living laboratories help Universities to adapt more swiftly to the needs of society? What are the examples in the Gulf region that we can build on? How do students and researchers experience University Living Labs? Can it be a model for the Gulf region?

In a study on University Living Labs and governance, case studies highlighted the value of a University living lab in terms of achieving research impact. The impact is broadly recognized as a strategic priority and responsibility for universities to demonstrate the social relevance of research and education and living labs provide spaces for researchers to meaningfully engage and partner with diverse stakeholders, develop a shared understanding of problems, and learn from real-world development and accelerated testing of solutions to global challenges. This impact agenda is a significant part of University communications, identity, and demonstration of leadership. Part of this value proposition is the opportunity to leverage research funding through grants and industry partnerships, and operational investment in University facilities (Hadfield et al. 2023). Living labs are not only a versatile tool to implement education for sustainability, contributing to the development of sustainability-related competencies among students, faculty, and staff, but they can also help Universities to transform their operational formats allowing them to gain a more relevant role in society and in the global process of transitioning towards a more sustainable world (Filho et al. 2022).

These benefits and opportunities are not without challenges. Living labs are difficult to organize and the coordination requires additional time and resources (Steen and Van Bueren 2017). In addition to these limitations, living lab projects might also face challenges associated with power issues between actors and end-user reluctance to engage (Hakkarainen and Sampsa 2013). Monitoring the works is another usual challenge—to make sure valuable and inefficient approaches are identified, as well as to facilitate replication and upscaling in different settings (Van Geenhuizen 2018).

Transforming a system requires significant energy and resources, usually over a considerable period. It also usually requires collective action among a wide group of partners who are willing to co-develop and commit to a collective vision. In some cases, transformation attempts can fail, particularly where there is strong inertia or opposition to change (Stibbe and Prescott 2020). Some of the challenges identified in the implementation of living labs refer to: the complexities in institutional administration, the tensions between different groups of interest that need to be addressed by enhanced communication, and the necessity to pay attention to the demand of using sustainability and innovation as a strategy in the operations of living labs (Filho et al. 2022).

3.4 Future Visions: Transforming the Campus and the Community

Water, energy, and food are necessary for the benefit of human well-being, poverty reduction, and sustainable development. Improved water, energy, and food security on a global level can be achieved through a nexus approach—an approach that integrates management and governance across sectors and scales. A recent conference panel in Muscat, Oman discussed the food-water-energy nexus in the Gulf region, bringing together speakers from academia (Sultan Qaboos University, Nizwa University, University of Vermont) and government (Oman Ministry of Agriculture, Ministry of Planning) and NGOs (Food4Humanity), looking at multiple scales, from micro-nutrients to community projects, to University campuses and large watersheds, and in different cultural and political contexts, including Oman, Qatar, Iran, and war-torn Yemen (Grichting 2022a, b). The session highlighted the complexity of addressing these "wicked" problems of systems and nexus thinking, and how our education and governance systems need to adapt to remove the barriers and silos that prevent these interdisciplinary and cross-cutting approaches. The nexus solutions must also address the quality, and not just the quantity of resources, be it water or food, as these have an important impact on human and ecosystem health. GIS and modeling tools can assist in integrating complex data and envision scenarios for more efficient management of resources, and collaboration between academic researchers with the government ministries managing food water and energy resources is important to develop new and practical solutions. Sea water should also be considered a water resource for its benefits of minerals (magnesium) and production of halophytes and sea-water pools could both save fresh water and provide much-needed minerals. Practical examples from war-torn Yemen (Food4Humanity) demonstrated the effectiveness and resilience of community-led projects to address food, water, and energy shortages and how they can improve the livelihoods and education of the population, especially women and youth. Finally, based on the presentation of the Qatar University Living Lab, discussions were led on creating a living laboratory for the food-water-energy nexus at Sultan Qaboos University as a way to engage multiple stakeholders and to create interdisciplinary collaborations.

Exploratory research highlights different approaches to living labs, and their potential to act as accelerators of sustainable practices and applied research on University campuses. To further develop the research and case studies, one would have to go beyond the classification of the living laboratory, to look at all Universities in the region that are developing initiatives and strategies that are similar to the living lab principles, or that are integrating certain aspects of a living lab, and to understand how these institutions understand and conceptualize sustainability and climate change. Can a common Gulf vision for sustainable development be transformed into a research and educational strategy for Gulf Universities, and can the living laboratory model contribute to creating the framework for these new models of Universities?

What will the global Universities of the future look like? And what about future Gulf Universities? A recent meeting to discuss trends in education and research in the Swiss University landscape concluded that interdisciplinary and creative research is essential for innovation, new professional fields call for new skills, and Universities need to better reflect this transformation by stepping up cooperation with stakeholders from society and business (SWITCH 2023). Living laboratories are part of these new interdisciplinary and multi-stakeholder research cultures, capable of addressing the multidisciplinary character of sustainability, in that they are solution-oriented and tailored to local realities, applicable to local policy-development and decision-making processes, and they engage local communities in defining the research agenda (Filho et al. 2017). Boundaries need to be better bridged by learning how to handle conflicts and deal with intermediation while respecting shared goals and interests and University living labs call for maintaining a solid relationship with cities and their actors. An evaluation framework is a key part of the working plan of living labs, where strong attention is required for boundary-spanning in evaluation, as living labs are increasingly applied in comprehensive multi-activity settings (Van Geenhuizen 2018).

Since the nineteenth century, the University has played a key role as a site of knowledge production and scientific and social innovation, and governments and other institutions routinely cite University publications as policy justification (Tegama and Fox 2023). Most Universities are still built and structured on patriarchal and colonial ideologies, but the knowledge and skills needed today are no longer aligned with these models. Future research will need better descriptions than were available to us about the "who," "what," and "where" of higher education and its part in social transformation as well as about the role of Universities in regional regeneration and their contribution to the creation of knowledge economies and fairer more socially inclusive societies (Brennan et al. 2004). Therefore, a paradigm shift in education is necessary for more inclusive and sustainable development requiring structural and ideological transformations in all our educational and governing institutions that can overcome disciplinary barriers as well as those related to race and gender, and those that separate man from nature and all living beings.

References

AFED (2016) Arab Environment: Sustainable Development In: Saab N, Sadik A (eds) A changing Arab climate. Annual report of Arab forum for environment and development, Beirut, Lebanon. Technical Publications

Bajgier SM, Maragah HD, Saccucci MS, Verzilli A, Prybutok VR (1991) Introducing students to community operations research by using a city neighborhood as a living laboratory. Oper Res 39(5):701–709

Bodin Ö (2017) Collaborative environmental governance: achieving collective action in social ecological systems. Science 357(6352):1114. https://doi.org/10.1126/science.aan1114

Brennan J, King R, Lebeau Y (2004) The role of universities in the transformation of societies: an international research project. Centre for Higher Education Research and Information/Association of Commonwealth Universities

Brookings (2023) The blockade on Qatar helped strengthen its economy, paving the way to stronger regional integration. https://www.brookings.edu/blog/order-from-chaos/2021/01/19/the-blockade-on-qatar-helped-strengthen-its-economy-paving-the-way-to-stronger-regional-integration/

European Network of Living Labs (2024) https://enoll.org/network/living-labs/

Evans J, Karvonen A (2013) Give me a laboratory and I will lower your carbon footprint!—urban laboratories and the governance of low-carbon futures. Int J Urban Reg Res 38(4):413–430. https://doi.org/10.1111/1468-2427.12077

Filho WL, Andrade GJ, Mifsud M, Pretorius R (2017) Universities as living labs for sustainable development: a global perspective. World Sustain Ser 26:8–15

Filho WL, Salvia AL, Pretorius RW, Brandli LL, Manolas E, Alves F, Azeiteiro UM, Rogers J, Shiel C, Do Paco A (2020) Universities as living labs for sustainable development. Springer, Berlin

Filho WL, Gokcin-Ozuyar P, Dinis M, Garcia-Alvarez M, Da Silva-Neiva S, Salvia A, Borsari B, Nita-Danila A, Vasconcelos C (2022) Living labs in the context of the UN sustainable development goals: state of the art. Sustain Sci 18:1240. https://doi.org/10.1007/s11625-022-01240-w

Global Forum/Water and Humanity (2022) The Muscat call 2022. https://www.waterandhumanity.com

Grichting A (2017) A productive permaculture campus in the desert: visions for Qatar University. Fut Food J Food Agricult Soc 5(1):21–33

Grichting A (2022a) Educational frameworks for designing regenerative food systems in the Arabian Gulf: the case of Qatar. In: Harriss H, Salama AM, Lara AG (eds) The Routledge companion to architectural pedagogies of the global south. Routledge, New York, pp 1–20

Grichting A (2022b) Report on the panel on water, food and energy nexus. Global Forum, Muscat, W-Hotel

Grichting A (2019) M-NEX Doha international workshop. In: Community stakeholder engagement and the urban farm (workshop report)

Grichting A (2020) Blue design for urban resilience in drylands: the case of Qatar. In: Nature driven urbanism, pp 175–208. https://doi.org/10.1007/978-3-030-26717-9_9

Hadfield P, Sharp D, Zarea M, Pigeon J, Peng X, Rye S, Raven R (2023) Governing University living labs for sustainable development: lessons from international case studies. Monash Sustainable Development Institute (Monash University). https://doi.org/10.26180/22138073

Hakkarainen L, Sampsa H (2013) How do we keep the living laboratory alive? Learning and conflicts in living lab collaboration. Technol Innov Manage Rev

Javatpoint (2024) Top 10 richest universities in the world. https://www.javatpoint.com/top-10-richest-universities-in-the-world

KAUST Living Laboratory (2024) https://sustainability.kaust.edu.sa/kaust-living-lab/

KAUST (2023) Saudi initiatives impact assessment. https://www.kaust.edu.sa/en/study/saudi-initiatives/impact-assessment

Koch N (2017) Green laboratories: University campuses as sustainability "exemplars" in the Arabian Peninsula. Soc Nat Resour 31:546. https://doi.org/10.1080/08941920.2017.1383546

König A (2013) Regenerative sustainable development of universities and cities: the role of living laboratories. Edward Elgar Publishing. https://doi.org/10.4337/9781781003640

Lupp G, Zingraff-Hamed A, Huang JJ, Oen A, Pauleit S (2021) Living labs: a concept for co-designing nature-based solutions. Sustainability 13(1):1–22. https://doi.org/10.3390/su13010188

Mitchell WJ (2003) Me++: the cyborg self and the networked city. MIT Press

M-NEX (2019a) Doha living lab for the food water energy nexus: program of the international workshop Qatar (A. Grichting, Ed.)

M-NEX (2019b) Moveable nexus: design-led urban food, energy and water management innovation in new boundary conditions of change. http://m-nex.net/about/

M-NEX (2024) Doha design workshop: building a food infrastructure—the Qatar experience. http://m-nex.net/m-nex-doha-design-workshop-february-2019building-a-food-infrastructure-the-qatar-experience/

RAMSAR (2023) Bio-cultural diversity. https://www.ramsar.org/activities/bio-cultural-diversity

Stibbe D, Prescott D (2020) The SDG partnership guidebook: a practical guide to building high-impact multi-stakeholder partnerships for the sustainable development goals. The Partnering Initiative/UNDESA. https://sdgs.un.org/publications/sdg-partnership-guidebook-24566

Steen K, van Bueren E (2017) Urban living labs: a living lab way of working

SWITCH (2023) Future university. https://www.switch.ch/about/future-University/

Tegama N, Fox A (2023) Ethics, the university, and society: toward a decolonial approach to research ethics. Alliance for African partnership perspectives, vol 2. Project MUSE, pp 75–91.https://muse.jhu.edu/article/911264

Times Higher Education (2024) World university rankings: Arabic university rankings 2021. https://www.timeshighereducation.com/world-university-rankings/2021/arabic-university-rankings

Torba Farms (2023) Meet the Qatari farmer growing medicinal plants in the desert. https://www.euronews.com/culture/2022/07/25/meet-the-qatari-farmer-growing-medicinal-plants-in-the-desert

UN Sustainable Development Goals (n.d.) Goal 17: partnerships for the goals. https://sdgs.un.org/goals/goal17

Van Geenhuizen M (2018) A framework for the evaluation of living labs as boundary spanners in innovation. Environ Plan C Polit Space 36(7):1280–1298. https://doi.org/10.1177/2399654417753623

Anna Grichting is an architect, urbanist and musician from Switzerland and holds a Doctor of Design from Harvard University. As an educator, researcher and international consultant, her major research interests lie in indigenous and interdisciplinary approaches to regenerative urbanism, nature-based design, blue design, food-water-energy nexus, participative and multistakeholder practices and border landscapes in conflict areas. She has taught and conducted research at Qatar University, University of Geneva and Harvard University. She was a fellow at MIT Jersualem 2050 and is currently a non-resident Senior Research Fellow at the University of Vermont.

4 Using Project-Based Learning to Nurture Sustainable Competencies Among University Students in Kuwait

Sosil Somokian

Abstract

This chapter investigates the critical challenges and opportunities in delivering Education for Sustainable Development (ESD) within Kuwait's higher education sector, contextualized by the national objectives of Kuwait Vision 2035. It presents a detailed case study analyzing the transformative learning experiences of 176 female students at Box Hill College Kuwait who engaged in Project-Based Learning (PBL) curricula designed around the Sustainable Development Goals (SDGs). The analysis reveals how PBL pedagogy significantly enhanced students' understanding of environmental issues, strengthened their sense of agency, and cultivated essential competencies in critical thinking, collaboration, and creativity. By demonstrating a practical model for effectively embedding ESD, the chapter argues that sustained, cross-curricular PBL initiatives are a powerful mechanism for building institutional capacity and fostering an ecologically conscious generation, offering valuable insights for policymakers and educators across the Gulf region and beyond.

Keywords

Education for sustainable development (ESD) · Higher education · Kuwait · Project-based learning (PBL) · Sustainable development goals (SDGs)

4.1 Introduction

In 2015, the United Nations adopted the Sustainable Development Goals (SDGs) with a 2030 deadline, emphasizing the need for a collective global effort to address challenges and build a sustainable future. Kuwait has aligned its efforts to achieve the SDGs outlined in the Strategic Cooperation Framework for 2020–2025 and to support Kuwait Vision 2035 through the National Development Plan of Kuwait (KNDP). Article 12 of the Paris Agreement highlights the importance of higher education in achieving SDGs, emphasizing the need to enhance climate change through education, training, public awareness, public participation and public access to information.

S. Somokian (✉)
Box Hill College of Kuwait, 29192, Exit 211, Abu Halifa, Al- Ahmadi Governorate, Kuwait City 54600, Kuwait
e-mail: s.somokian@bhck.edu.kw

K. Al'Abri and E. Buckner (eds.), *Higher Education and Sustainable Development in the Arab Gulf States*, Sustainable Development Goals Series, https://doi.org/10.1007/978-3-032-19559-3_4

University students should have the information, critical literacy, social connections, appreciation for diversity, moral responsibility, and civic engagement necessary to address today's pressing issues (UNESCO 2017). Universities are expected to equip the students with the knowledge and the skills needed to address environmental problems (SDG 13), which often have economic causes (SDG 8) and social consequences (SDG 5 and 10).

The complexity and interconnectedness of the SDGs require an interdisciplinary approach to education, necessitating collaboration by multiple university departments to equip students with the knowledge and skills needed to comprehend and tackle complex issues like poverty, climate change, and inequality (Tejedor, Segalas, and Rosas-Casals, 2018). However, many universities face challenges, including limited resources, a lack of faculty training on sustainable development, and a siloed culture that hinders collaboration (Leal Filho et al. 2019). Additionally, the rapidly evolving field of sustainable development requires constant updates to curricula, which universities struggle to keep up with. Despite these challenges, universities in the GCC, especially in disciplines like engineering, business, IT, and medicine, are gradually incorporating SDGs into their curricula and embracing interdisciplinary approaches to educate students and strengthen capacity for the future. Besides making Education for Sustainable Development (ESD) a core component of all educational programs, providing holistic education, helping learners gain an understanding of multiple fields, and preparing a generation of change agents capable of transforming their respective environments requires reforms in teaching and learning methods.

Traditional teaching methods are insufficient and often fail to integrate the social, economic, environmental, and cultural dimensions of SDGs (Scarff Seatter and Ceulemans 2017). UNESCO advocates for a focus on *learning* rather than *teaching* in Higher Education for Sustainable Development (HESD). A student-centered approach like Project-Based Learning (PBL) is more effective as it promotes active engagement, deeper understanding of environmental problems, critical thinking, collaboration, and real-world problem-solving. Such methods foster innovation, creativity, and the application of knowledge to sustainability challenges.

Higher education institutions in GCC have been implementing experiential and Project-Based Learning (PBL) in various fields in the last two decades to connect students to the real world and prepare skilled workers capable of applying course knowledge to solve real-life problems. At Box Hill College Kuwait, the faculty of the Intensive English Program and the School of Art and Design designed PBL courses to help learners gain key sustainability competencies and develop the skills and knowledge needed to become change agents in the Kuwaiti community.

This two-year exploratory case study at Box Hill College Kuwait examined students' perceptions of Project-Based Learning (PBL) in terms of its opportunities, challenges, and impact on developing employability skills, their understanding of SDGs, and whether learners saw any role for themselves in achieving Kuwait's 2030 vision and goals. Pretests and post-tests assessed students' knowledge of climate change, recycling, and sustainability. Data was collected using surveys, semi-structured interviews, self-assessments, and student journals, and analyzed through SPSS and inductive methods. The study tracked student experiences and learning through orientation and project implementation phases, while follow-up interviews were conducted to understand learners' evolving perceptions of PBL after completing multiple courses.

Findings indicated that after completing any one of the PBL courses, almost all learners recognized the link between environmental conditions and health problems, and the majority viewed it as an urgent issue. Although only half of those surveyed believed their personal behavior could make a difference in sustainability matters, all learners enrolled in at least one PBL course felt empowered to act. Regarding skill development, at least two-thirds of learners reported an increase in sustainability competencies, particularly in strategic thinking, with no reports of a decrease in skills. The findings of this study provided useful feedback to educators and policymakers involved in this initiative, highlighting the strengths of PBL and areas that needed improvement. This chapter examines the efficiency of Project-Based Learning, addressing specifically three questions: *its usefulness*—To what extent do students find Project-Based Learning (PBL) useful for learning about environmental responsibility? *Its impact on learners' skills and understanding*—How did PBL impact on students' understanding of environmental problems and their development of interpersonal, cognitive, and professional skills? And *the extent of its effectiveness*—Was the improvement in student outcomes significantly greater for those who took three PBL courses compared to those who took only one.

4.2 Literature Review

Project-Based Learning (PBL), rooted in John Dewey's "learning by doing" theory, emphasizes experiential learning and self-reflection. Initially used in medical education, PBL is now applied across various fields. It is student-centered, context-specific, and promotes active participation, collaboration, and real-world problem-solving, fostering critical thinking and the application of knowledge to authentic challenges.

Research on Project-Based Learning (PBL) has highlighted its effectiveness in enhancing student achievement and identified key factors for its success, such as learner characteristics, teacher readiness, and the role of technology. UNESCO encourages the integration of Education for Sustainable Development (ESD) into university curricula to foster students as change agents for sustainability. Among real-world learning methods, PBL is considered transformative in ESD, promoting active learning and collaborative problem-solving. Its flexibility supports "liquid learning," (Savin-Baden 2014, p. 197 & 210) where knowledge continuously evolves across disciplines, helping students develop adaptive skills for uncertain futures. PBL supports the development of sustainability competencies, including systems thinking, critical thinking, and collaborative problem-solving, making it valuable for interdisciplinary curricula.

Aligned with ESD's focus on holistic, critical, and action-oriented learning, PBL transforms educators into facilitators who guide students in collaborative problem-solving. Scholars like Tejedor and Lozano emphasize PBL's effectiveness in delivering SDGs and fostering sustainability competencies in higher education. "University teaching is often separated from its surroundings, and many professors are unprepared to teach such abilities (Tejedor et al. 2019). Lozano believes that Project-Based Learning is a highly effective pedagogical approach and if structured carefully, both PBL and problem-based learning can develop a wide range of sustainability competencies (2017). Studies in Western societies demonstrate PBL's positive impact on capacity building and its effectiveness in supporting ESD in higher education."

Several studies have been carried out in GCC to assess the effectiveness of Project-Based Learning (PBL) in higher education institutions within the GCC. Trevor P. Ray's research on Emirati students in Business and IT found that 4% of students found PBL more engaging, while 83% reported increased motivation to learn (Ray 2017). Similarly, a study at Qatar University involving 90 students from the Electrical Engineering Department showed strong positive feedback on PBL. Ninety-seven percent of learners found the courses (ELEC333 and ELEC371) more interesting, 94% believed it enhanced their understanding of theory, 83% improved their knowledge of electronics and sensors, 88% felt they gained additional skills, 98% saw an improvement in real-life problem-solving, 72% improved teamwork, and 92% reported enhanced creativity. Additionally, 92% of students expressed interest in taking more PBL courses (Khandakar and Ayari 2020).

In an action research study at Bahrain Polytechnic, Huijser and Hasan explored how PBL enhanced Bahraini learners' employability, focusing on entrepreneurial and innovative skills. The pilot study, based on teacher interviews and classroom observations, found that PBL helped develop graduates' soft skills and better prepared them for the job market. The study's conclusions were drawn from quantitative data, specifically learners' grades on course projects. In a study in the UAE, Mazrouni identified challenges in implementing PBL in the moral education curriculum, focusing on project design and assessment methods. Mazrouni suggested adopting PBL to enhance student engagement in community development. The study highlighted that GCC university students have a positive attitude toward PBL and value its role in developing employability skills, though they often lack basic knowledge of sustainability issues.

Radwan and Khalil (2021) surveyed 200 students at the University of Sharjah to assess their knowledge and attitudes toward sustainability. The results showed that students had strong knowledge of UOS sustainability programs and supported initiatives like responsible water use and eco-friendly products. However, they opposed limiting car usage on campus. The study highlights the need to bridge the gap between sustainability knowledge and action and suggests improvements for sustainability communication, in line with the UAE's 2030 Agenda for Sustainable Development.

A large-scale study by Al Saati, S. Al Nakla, and D. Al Nakla surveyed 500 students from seven universities in Saudi Arabia's Eastern Province, revealing a lack of sustainability knowledge. Many students could not identify common recyclables or renewable resources and showed little concern for water and energy conservation, although some expressed interest. Two-thirds had never participated in recycling, with 65% blaming their schools for not providing recycling bins. The study found that most students were unaware of the importance of sustainability, and only a few showed genuine concerns for environmental issues. The researchers concluded that sustainability awareness should be promoted through media, government, and university efforts, with all stakeholders involved in developing policies to integrate and apply sustainable practices.

A study of 152 students at the College of Architecture and Planning at the University of Dammam found that while students were aware of campus sustainability issues, they were unwilling to engage in sustainability initiatives. Despite some efforts in transportation, energy, and water conservation, courses and projects had limited focus on sustainability. The study emphasizes the need to integrate sustainability into campus operations and train students to promote environmental sustainability in Saudi Arabia and the Middle East.

A study of eight architecture programs in the UAE found that while sustainability is recognized as a core topic in official literature, its integration into courses is insufficient. The study highlighted a gap between studio and theory-based courses, with sustainability rarely introduced in studio classes. This calls for a shift in architectural education to fully integrate sustainability into curricula and teaching practices, aligning with sustainable city planning and development goals in the UAE.

In recent decades, higher education institutions in Kuwait have used PBL to teach specific courses, but not as an integrated educational approach. Gulf University for Science and Technology, Australian University of Kuwait, and Kuwait University have attempted to implement PBL in fields like engineering, marketing, management, and business studies. However, only Australian University of Kuwait offers a full program based on PBL. Small-scale comparative studies in Kuwait aim to compare PBL with traditional teaching methods, measure student and teacher attitudes, and assess PBL's role in sustainable development education. In conclusion, higher education institutions can play a significant role in achieving environmental sustainability both on campus and in the wider community. Universities in the GCC must adopt Education for Sustainable Development (ESD), embed SDGs into their curricula, and embrace student-centered approaches like PBL to prepare students for their roles in achieving sustainability goals.

4.3 Research Objectives and Questions

While cultural factors are known to influence perceptions of sustainability, it is not well understood how pedagogical approaches like PBL are perceived by students in this specific context. This study directly investigates the extent to which students in Kuwait find PBL a useful tool for learning about environmental responsibility, thereby incorporating student voice into the design of locally relevant ESD. Moreover, there is a significant lack of empirical research measuring the multi-faceted impact of PBL in GCC universities. The impact of PBL on building the specific suite of skills necessary for sustainability—including cognitive, interpersonal, and professional competencies—remains largely unmeasured. Furthermore, the potential of technology and social media to enhance PBL's engagement and interactivity, despite student affinity for these tools, is consistently overlooked. This study fills this gap by providing empirical data on how PBL impacts a comprehensive range of skills and knowledge. Furthermore, existing research on ESD in the region predominantly focuses on short-term outcomes from single courses. A major gap exists in understanding the long-term effects and cumulative impact of sustained intervention. This study directly investigates this by comparing outcomes between students enrolled in a single PBL course versus three courses, providing crucial insights into how to design multi-semester, interdisciplinary curricula for maximum effect in building sustainability knowledge and shifting student attitudes.

Specifically, the study focuses on three key questions. First, what are teachers' and students' views regarding the effectiveness of Project-Based Learning (PBL) approach in achieving course objectives and developing employability skills? Second, what are the challenges faced by students and teachers in achieving SDGs through PBL. Third, how do the learning outcomes of students enrolled in one PBL course compare to those enrolled in three PBL courses, with a focus on developing employability skills and learners' knowledge of sustainability.

4.4 Data and Methods

This two-year exploratory case study investigated students' perceptions regarding PBL as an approach to teaching and learning, the opportunities it created, the challenges it presented, and the impact these courses had on developing their employability skills, understanding the importance of SDGs and instigating a sense of urgency toward achieving SDGs for Kuwait 2030.

The study was conducted at Box Hill College Kuwait where English is the medium of instruction. Pretests and post-tests were done to assess the depth of undergraduate students' knowledge about climate change, recycling, and sustainability before and after completion of the course. Data were analyzed using SPSS. Surveys, semi-structured interviews, and self-assessments were used to collect data prior, during, and after the implementation of the projects. Students were asked to keep journals to record and share details of their learning journey with the interviewers.

The results of student semi-structured interviews and surveys were analyzed using an inductive method. The responses of the students were coded in an iterative manner, and the emerging patterns were tallied. Analytical and evaluative questions were used to collect data from learners during the orientation and implementation phases of the projects. Analytical questions were used to understand details of what, when, and how of the learning process, while evaluative questions were asked to understand the students' reasoning about situations and outcomes (the why). Surveys were carried out at the end of the project to gauge students' sentiment about PBL as a method of teaching and learning. Follow-up interviews were done with the 37 learners enrolled in all three PBL courses upon their completion of each course, to understand how they experienced PBL the second and third time.

A total of 176 Kuwaiti female students were recruited for this study. They were in preparatory, first year, and second year of their university studies. 26 learners were enrolled only in 2D-3D design PBL version (first year course in Art and Design). 113 learners were enrolled only in CRW2 (Critical Reading and Writing 2, a second-year general education course mandatory for all BHCK students). 37 learners were enrolled in all three courses EAP (English for academic purposes), CRW2 and 2D-3D design over consecutive semesters.

The study was carried out during Fall 2021, Spring 2022, and Fall 2022. The five teachers facilitating the three courses had gone through a teacher training course during Spring and Summer 2021 to design and facilitate the PBL version of their courses and had an agreed-upon framework for facilitating the SDGs and the core concepts of their units. The cohort of 37 students was interviewed and surveyed in Fall 2021, Spring 2022, and Fall 2022. This group was enrolled in three PBL courses in consecutive semesters. While the 113 students enrolled in CRW2, and the 26 students enrolled in 2D-3D design were interviewed and surveyed for the single PBL course they studied in Spring 2022 and Fall 2022. EAP courses in PBL version were offered in Fall 2021 and Spring 2022, CRW2 in PBL version was taught in Spring 2022 and Fall 2022, while 2D-3D design in PBL version was offered in Fall 2021 and Spring 2022.

Of the 176 students 26 were enrolled in 2D-3D design; 113 were enrolled in Critical Reading and Writing (CRW2- a second-year course) and only 37 were enrolled in three of the 4 PBL courses—EAP1, EAP2, CRW2 and 2D-3D design, over a period of 3 semesters. While the four PBL courses shared the same framework and had similar objectives, they were essentially distinct in character and activities and could not be directly compared. The main goal of this study is to shed light on students' perceptions of the impact of PBL on their learning about environmental responsibility, encouraging them to change their behavior, take action to raise awareness among their peers and community on issues such as global warming, and understand the importance of recycling, keeping the desert clean, and striving to create a greener campus.

The cohort of 37 female students enrolled in three PBL courses completed the project titled "Keep Our Beautiful Desert Clean." The cohort of 26 students enrolled in the "2D-3D Design" course used recycled water bottles to create 2D-3D models and leveraged social media to raise awareness among their peers about the importance of recycling. Meanwhile, 113 students enrolled in the "Critical Reading and Writing 2" (CRW2) course designed animated stories to help their peers understand global warming and recognize their role in creating sustainable environments. This study summarizes the findings of surveys and interviews conducted with students and teachers involved in different courses to assess the effectiveness of Project-Based Learning (PBL) in achieving the Sustainable Development Goals (SDGs), to highlight the strengths of the method, and to propose strategies for enhancing the development of sustainability competencies among university students in Kuwait.

4.4.1 Data Collection

Four PBL courses were designed and implemented in classroom settings. The preparatory year English course for intermediate and advanced levels (EAP1 and EAP 2) taught in 12 sessions of 1.5 h on weekly basis, of which 6 h were dedicated to PBL. CRW2 and D2-D3 Design courses were carried out over 3 sessions of 2 h on a weekly basis. All courses ran for 16 weeks (full semester). Group sizes varied between 17 and 29. The EAP groups were smaller while the CRW2 groups were larger. In larger groups, a tutor assisted teachers. Students were divided into subgroups of four or five and asked to work on their projects. Both the tutor and the teacher assumed the role of the facilitator after the introduction phase (set-up and presentation of the driving question).

Multiple forms of data were collected, including a test with multiple-choice questions about sustainability and environmental topics, teacher observation forms, student journals and semi-structured interviews.

First, analytical and evaluative questions were used to collect data from learners during the orientation (Weeks 2–4) and implementation phases (Weeks 5–16) of the projects. Analytical questions were used to understand the "what, when, and how" of the learning process while evaluative questions were used to understand the students' reasoning about situations and outcomes, the "why." Data collection instruments were used pre-intervention (Week 1—survey, self-assessment, and multiple-choice test of knowledge), post-intervention (Week 17 survey, self-assessment, multiple-choice test, and interviews) and during intervention (students' journal entries, teacher observations and part of the semi-structured interviews during Weeks 2—16). The multiple-choice test and competence-oriented self-assessment were used before and after intervention (PBL course) to establish a baseline and note changes in the level of knowledge and sustainability skills due to the intervention. TheThis chapter examines pretests and post-tests consisted of 20 multiple-choice questions covering factors leading to global warming, the consequences of climate change on the environment, health, and the economy, the rules of recycling plastic, paper, and hazardous materials, and factors supporting sustainability.

Semi-structured interview was done in two steps, some questions were used to collect data about the challenges of the process as students were going through the course, and some questions were reserved for after submission of the final product (post-intervention) to gauge their attitude and perception of the whole process. The group of students (37) who studied three PBL courses was interviewed during and after completing the first PBL course (EAP) and had two follow-up interviews after completing CRW2 and 2D-3D Design courses.

The study utilized different instruments through various stages of intervention to measure students' learning outcomes and experiences with Project-Based Learning (PBL). A post-intervention survey gauged students' perceptions of PBL's usefulness in learning about environmental responsibility. Quizzes administered before and after the intervention assessed students' knowledge of SDGs, while competence-oriented self-assessments measured changes in interpersonal and cognitive skills. Semi-structured interviews conducted during and after the intervention explored students' experiences with PBL, focusing on challenges, highlights, and how it enhanced their understanding of the issues, teamwork, communication, and problem-solving abilities. Follow-up interviews with 37 students tracked changes in their perceptions of PBL's role in achieving SDGs after completing three PBL courses.

In addition, teacher observations were collected to record shifts in students' understanding of SDGs, sustainability skills, and attitudes toward their role in achieving these goals. Observations were conducted four times, every two weeks, throughout the task. Teachers assessed students' collaboration, initiative, research, knowledge sharing, and contribution to the final presentation. These observations also helped identify areas where students needed extra support, prompting the involvement of subject matter experts for guidance. To record the results of their observations, teachers used a 5-point Likert scale (1–5: Poor, Below Expectations, Meets Expectations, Exceeds Expectations, and Exceptional).

There are a number of limitations with the study. First, the subjects recruited for this study were all female, which limits the generalizability of the findings, as the results and female learners' feedback or perceptions may not reflect the experiences or learning outcomes of wider student population. Moreover, gender-related cultural and social factors may have influenced the female participants' perceptions and engagement, making the results specific to this group. Additionally, the absence of male students makes it difficult to assess how gender diversity might have affected group dynamics and learning outcomes in the courses. Another limitation is that all the PBL courses were part of the program requirements, meaning learners had no choice in participation. It would have been ideal to run the same course simultaneously in two different modes (PBL and traditional) and randomly assign learners to each class, allowing for comparison in learning outcomes and the impact of PBL as a pedagogy on students' acquisition of SDGs.

Additionally, group sizes in the three PBL courses varied from 17 to 29, and some groups may have been stronger than others, with group dynamics potentially influencing students' engagement and learning.

Furthermore, the three projects differed in complexity, which could have affected learners' overall attitudes toward PBL, their engagement, and their enthusiasm to share their thoughts and feelings during the interviews.

Moreover, each PBL course was taught by a different instructor, and since teachers' skills and experience are key to the success of PBL, this could have influenced the results, even though all instructors participated in professional development (PD) courses to learn about PBL and its implementation. With the help of specialized trainers, teachers prepared the PBL version of their courses, including course content, lesson plans, rubrics, observation templates, and assessment criteria. Despite the training, individual differences among teachers persisted.

Another factor to consider is that learners were at different stages of their studies when enrolled in the PBL courses, which may have affected their performance, as their interpersonal and cognitive skills likely evolved over time.

The tools used for data collection also limit the reliability and generalizability of the study due to the sample size and multiple variables (three PBL courses). However, this was compensated for by the large amount of detailed data collected through semi-structured interviews over two years, which allowed for the emergence of themes. The data collected was later cross-checked with the teachers and learners.

4.5 Results

To understand undergraduate students' sense of environmental responsibility, it was essential to assess their knowledge of climate change, recycling, and sustainability. As shown in Table 4.1, the results indicated that most students had limited awareness of these topics prior to enrolling in any of the PBL courses but gained significant knowledge after completing one or more of the PBL courses. The group averages were as follows:

Upon completing any one of the PBL courses (all of which included some content on the SDGs), almost all learners (98%) stated that there was a direct relationship between environmental problems and health and well-being. Additionally, 86% of learners believed the issue was urgent. Interestingly, while only 54% believed their own behavior could make an impact, 94% of those enrolled in three PBL courses (35 out of 37 students) believed in the importance of acting now and in their ability to make a difference. This was measured using surveys and cross-checked with detailed questions in semi-structured interviews. Students were surveyed both before and after completing a PBL course, as well as after completing three PBL courses. Regarding the impact of enrollment in PBL courses on their interpersonal, cognitive, and professional skills, as shown in Table 4.2, the following results were recorded: 95 out of 176 students reported a positive increase in at least one of the four sustainability competencies over the duration of their project, with future thinking, strategic thinking, and values thinking being the most frequently reported. "Decrease" was not reported. Twelve students did not return the competence-oriented self-assessment forms. The results of 164 students' self-assessment forms are summarized in the table.

As shown in Tables 4.3–4.6, teachers reported a steady increase in learners' interpersonal skills, cognitive competencies, and professional conduct upon completion of a PBL courses. Table 4.3 shows student competencies for students in EAP, Table 4.4 for students in 2D-3D, Table 4.4 for students in CRW2, and Table 4.6 for students in all three PBL courses. For interpersonal skills, they observed students' competencies in teamwork, leadership, collaboration, negotiation, and argumentation. For cognitive skills, they focused on learners' abilities to research and find information, analyze problems, make decisions, find solutions, and think critically and creatively (e.g., the creative use of social media to influence peers and raise awareness). For professionalism, teachers observed students' punctuality, preparedness before class, note-taking skills, and oral and

Table 4.1 Knowledge of climate change, recycling and sustainability (20 multiple choice quest)

Course	Pretest (%)	Post-test (%)	Difference(%)
EAP	10	65	+ 55
2D-3D	35	75	+ 40
CRW2	45	80	+ 35
EAP + CRW3 + 2D-3D	10	95	+ 85

Table 4.2 Students self-reported positive change in sustainability competencies (SC)

Course	Number of students	Systems thinking	Future thinking	Values thinking	Strategic thinking
EAP	37	17	19	29	24
2D-3D	26	16	15	23	22
CRW2	114	68	97	110	95
EAP + CRW2 + 2D-3D	148	128	126	132	125

written communication (emails, meeting minutes, reports). They also noted students' ability to share the workload and take responsibility for shortcomings Table 4.5.

The change in all competencies was significant when students were enrolled in three PBL courses compared to a single PBL course.

4.5.1 Student Assessments

Students were asked to journal their progress, documenting challenges and reflecting on their learning throughout the PBL course. They recorded questions, potential resources, and new insights, sharing their understanding and developing problem-solution plans through group discussions and mind maps. Weekly "I enjoyed—I hated" memos allowed for personal reflections, which were valuable for teachers in providing support. While most students initially disliked journaling, they later acknowledged its value in tracking progress and retaining ideas. Students utilized various tools, including mobile phones, iPads, and AutoCAD, for tasks like taking photos, recording interviews, and drafting prototypes. They also communicated via WhatsApp and email. The main challenges in the initial stages included limited knowledge of English, of ecosystems and human behavior, and the difficulty of identifying reliable resources. Critical thinking was needed to evaluate the information they found. Gradually, learners developed metacognitive skills and became more cognizant of biased sources. The final challenge was bridging theory with real-world applications, especially in the Kuwaiti context. Students explored global initiatives, piloted local projects, and adapted their solutions through debriefing sessions and facilitator discussions. Excerpts from student interviews on the role of PBL in enhancing their SDG knowledge and sustainability competencies are provided below.

> The best thing about this course was that **we** decided on our homework, **we** chose the experts to interview. We learned how to conduct an interview professionally and wrote the questions we wanted to ask. (EAP)
>
> This course taught me time management. I felt the pressure of deadlines, but it was ok, because I was solving an important problem, and that motivated me a lot. (CRW2)
>
> In the beginning, I did not like the team meetings, it was a waste of time. I wanted to work on my own and get my grade. But after a few weeks, I needed my friends' feedback. They were helpful, they made me rethink and replan." (EAP)
>
> I felt a lot of pressure because my team's success in the awareness campaign depended on the quality of my photos, and I didn't know how to make good photos. I had to search for famous photographers, check their and the posters of Greenpeace for hours, to understand what makes a good photo. I even watched online tutorials to learn how to position items to get the best shots of recycled items and make them tell a story. (2D-3D)
>
> "I am proud that I took all three courses where I learned through hands-on projects." It helped me become a much better student—I am more organized and strategic with my time now. I learned to read and analyze, and think about problems from different perspectives. I made a habit of asking myself, is there something I'm forgetting here?
>
> In these three courses, I read a lot about SDGs and watched videos about problems like poverty, inequality, migration that I hadn't heard about, I wish all our courses were projects. Because they help us think big and we study about things that matter and learn skills that help us solve problems.
>
> At first, I was scared to do public presentations, but after three PBL courses I look forward to presenting. These courses helped me become a confident speaker.

Table 4.3 Teacher observation of students' competencies in EAP

Competencies	Occ. 1	Occ. 2	Occ. 3	Occ. 4
Interpersonal	2.1	2.5	3.1	3.5
Cognitive	1.4	1.8	2.1	2.2
Professional behavior	1.2	1.6	2.2	3.6

Table 4.4 Teacher observation of students' competencies in 2D-3D design

Competencies	Occ. 1	Occ. 2	Occ. 3	Occ. 4
Interpersonal	1.2	1.8	2.1	2.3
Cognitive	1.5	1.8	2.2	2.6
Professional behavior	2.2	2.5	2.5	2.6

Table 4.5 Teacher observation of students' competencies in CRW2

Competencies	Occ. 1	Occ. 2	Occ. 3	Occ. 4
Interpersonal	1.1	1.4	1.5	1.7
Cognitive	2.1	2.3	2.2	2.8
Professional behavior	2.3	2.8	3.1	3.4

Table 4.6 Teacher observation of students' competencies after completing three PBL courses

Competencies	Occ. 1	Occ. 2	Occ. 3	Occ. 4
Interpersonal	3.2	2.5	3.1	3.5
Cognitive	2.2	2.8	3.1	3.2
Professional behavior	3.4	3.6	3.8	3.9

Speaking of the challenges of PBL, students mentioned time, lack of cooperation among peers, and managing the workload. "Time was always short, we had too many deadlines, peers did not always contribute or were not punctual in doing their part of the project." Students reported a lack of motivation in tackling complex aspects of their projects, particularly during the initial research phase when they struggled to find Arabic-language resources. A common complaint was uncertainty about how their work would be assessed, with some students expressing difficulty adapting to the new evaluation system. The following excerpts from student interviews highlight the challenges faced in PBL courses and their views on achieving Kuwait's 2030 SDGs.

> There was no time to do everything we wanted to do. At first, we were struggling with the sources that the teachers recommended to read or watch, they were all in advanced English. We had to translate them and summarize the main ideas, to prepare for the briefings (EAP).
>
> The driving question was tough. The stories we read about successful recycling campaigns were not applicable to Kuwait. Convincing others about the importance of saving water or recycling plastic bottles proved to be a huge challenge. They didn't see the point and were not interested. So, we had to be creative, to get the message across. (CRW2)
>
> For me the most difficult part was not knowing how to study to get good grades. I am an A-student, and grades are important to me. I didn't like that my grade depended on the final project, which was teamwork, as I didn't understand how the project would be evaluated. I needed teachers' guidance to understand the standards and plan the steps. (CRW2)

In the three PBL courses (EAP, CRW, and 2D-3D design), teachers assessed students through observations, journal entries, exams, and final presentations, giving equal weight to progress and final outcomes. Sustainability comprehension and its application were evaluated through interviews, which also gathered student feedback on their PBL experience. Oral presentations allowed students to showcase their work, practice persuasion, and encourage others to support sustainability initiatives. PBL's multi-component assessment provided instructors with insights into how students learned about sustainability in a project-based context.

4.6 Discussion and Conclusion

This study confirms and extends existing PBL literature by providing concrete evidence from the GCC context. The results align with global and regional findings, demonstrating that PBL enhances student engagement, motivation, and critical skills such as collaboration, creativity, and problem-solving. Crucially, however, this study addresses a central gap in GCC literature: the disconnect between sustainability knowledge and action. It shows that a well-structured PBL intervention can effectively bridge this gap by empowering students to become change agents. Learners reported feeling motivated by hands-on, real-world problem-solving, which led to improved attendance and punctuality compared to lecture-based courses. Furthermore, they developed stronger communication, critical thinking, and collaborative skills through designing awareness campaigns on issues like plastic reduction and water conservation. By linking learning directly to national initiatives such as Kuwait Vision 2035, this study offers a replicable model for how PBL can foster both practical skills and actionable environmental agency in GCC higher education.

The findings highlight the need for universities to intentionally integrate Project-Based Learning (PBL) into sustainability education as a structured component of the curriculum, as PBL provides students with hands-on opportunities to apply theoretical knowledge to pressing sustainability challenges, which not only deepens their conceptual understanding of the Sustainable Development Goals (SDGs) but also strengthens critical skills such as problem-solving, collaboration, and innovation. To translate these insights into practice, universities should embed PBL into core and elective courses by aligning projects with specific SDGs, ensuring that students engage with a variety of sustainability topics across their academic journey. Also, they should partner with local communities, NGOs, and industries so that student projects address authentic sustainability issues, thereby producing tangible social and environmental impact. Universities should also assess transversal skills alongside academic performance, incorporating evaluation of teamwork, creativity, and communication into grading criteria to better prepare students for sustainability-related careers. The findings also suggest that introducing longitudinal PBL experiences by spreading projects across multiple semesters enables students to revisit, refine, and expand their work in ways that mirror real-world sustainability initiatives. Finally, the findings indicate that PBL allows for interdisciplinary collaboration, bringing together learners from diverse fields—engineering, business, health, and social sciences—who together can co-develop solutions that reflect the complexity of sustainability challenges.

Upon reviewing the interview transcripts, student journals, teacher observation reports, and grades attained by students at various stages of the project and for the final product, clear patterns emerged. These findings contribute to a deeper understanding of the efficacy of Project-Based Learning (PBL) in the context of Education for Sustainable Development (ESD). The study found that students were more motivated and actively participated in class due to the flexibility to make decisions about their own learning. As a result, PBL's success can be directly attributed to the autonomy and freedom it provides. Additionally, the students' perception of working on authentic, real-life problems significantly enhanced their learning motivation and increased their level of engagement. Their understanding of the real-world significance and application of their work was a key factor in the success of PBL in the Kuwaiti context. A sense of personal responsibility for resolving complex issues, such as climate change and global warming, was positively correlated with Kuwaiti students' commitment to the project, the effort they invested, the progress they made, the grades they achieved, and their overall satisfaction with PBL. The students' learning experiences and progress were positively impacted by the advice and guidance provided by instructors and subject matter experts (SMEs). With appropriate scaffolding, students were able to learn more efficiently and achieve better results. The students' overall PBL experience was negatively affected by their inability to manage time effectively and communicate with classmates. They expressed frustration with "tight" deadlines and disappointment when teammates "did not pull their weight." Investing time and effort in developing key skills such as time management and communication prior to offering the PBL course would significantly improve students' ability to handle PBL challenges. It was noted that students, particularly those in the preparatory year, required additional time and support to understand and deconstruct the guiding question and the assessment rubric criteria. This was especially true for students who had not previously enrolled in PBL courses and for whom PBL represented a completely new and unfamiliar learning method. Combined, these identified patterns highlight the strengths and weaknesses of PBL implementation in the context of ESD. Educators and policymakers may benefit from a deeper understanding of these elements as they work to refine PBL as a method of delivery, aiming to enhance Kuwaiti students' engagement, motivation, and learning outcomes.

This study showed that PBL enhanced students' sustainability competencies by providing autonomy and decision-making flexibility, keeping them engaged and motivated. Working on real-world problems helped students understand their role in solving environmental issues and addressing climate change. They valued the opportunity to collaborate with peers, teachers, and experts while developing key skills and strategic thinking. However, challenges such as time management, communication, and uneven team contributions emerged early on, particularly for preparatory year students. Adequate preparation in the first phase, including developing skills like time management and negotiation, is essential for effective PBL. Balancing didactic instruction with inquiry methods fosters cognitive development and SDG knowledge. Overall, PBL proved effective in enhancing sustainability knowledge, problem-solving, and strategic thinking, even without a broad interdisciplinary approach. Enrolling in PBL courses with embedded SDGs helped students better understand sustainability and their role in achieving their country's development goals.

References

Khandakar A, Ayari MA (2020) Automatic and reliable leaf disease detection using deep learning techniques. Appl Syst Innov 3(2):20. https://doi.org/10.3390/asi3020020

Leal Filho W, Shiel C, Paço A, Mifsud M, Veiga Ávila L, Londero Brandli L et al (2019) Sustainable development goals and sustainability teaching at universities: falling behind or getting ahead of the pack. J Clean Prod 232:285–294. https://doi.org/10.1016/j.jclepro.2019.05.309

Lozano R, Ceulemans K, Seatter CS, Schneider L (2017) A review of research on higher education and sustainable development: 2005–2015. J Clean Prod 164:728–742. https://doi.org/10.1016/j.jclepro.2017.06.076

Radwan AF, Khalil EMAS (2021) Knowledge, attitude and practice toward sustainability among university students in UAE. Int J Sustain High Educ 22(5):964–981. https://doi.org/10.1108/IJSHE-06-2020-0229

Ray TP (2017) Emirati student and expatriate teacher views on problem-based learning in UAE education courses (Doctor of Education dissertation, University of Southern Queensland). https://doi.org/10.26192/5c05c3e0d30c9

Savin-Baden M (2014) Problem-based learning: new constellations for the 21st century. J Excel CollE Teach 25(3–4):197–219

Scarff Seatter C, Ceulemans K (2017) Teaching sustainability in higher education: pedagogical styles that make a difference. Can J High Educ 47(2):47–70. https://doi.org/10.47678/cjhe.v47i2.186284

Tejedor G, Segalàs J, Rosas-Casals M (2018) Transdisciplinarity in higher education for sustainability: How discourses are approached in engineering education. J Clean Prod 175:29–37. https://doi.org/10.1016/j.jclepro.2017.11.085

Tejedor G, Rosas-Casals M, Segalàs J (2019) Patterns and trends in engineering education in sustainability: a vision from relevant journals in the field. Int J Sustain High Educ 20(2):360–377. https://doi.org/10.1108/IJSHE-06-2018-0154

United Nations Educational, Scientific and Cultural Organization (UNESCO) (2017) Programme and meeting document. https://unesdoc.unesco.org/ark:/48223/pf0000261971

Sosil Somokian is the academic manager of the Intensive English Program at Box Hill College Kuwait. She holds Master of Arts degrees in Education and Speech and Language Pathology. Her research interests include education for sustainable development, innovative education for capacity building, the use of AI in classrooms, the efficiency of project-based learning, and problem-based learning, second language acquisition and learning difficulties.

Defining Education for Sustainable Development in the GCC: The Need for More Holistic Perspectives and Approaches

5

Hajar Mahfoodh and Hessa AlAtawi

Abstract

Many Higher Education (HE) academic bodies have produced policies that claim to integrate Education for Sustainable Development (ESD) in the Cooperation Council for the Arab States of the Gulf (GCC). Despite efforts to adopt Sustainable Development Goals (SDGs) in HE curricula, ESD has had difficulty being established effectively. This chapter explores the factors leading to misinterpretations and the current situation of ESD implementation in the HE bodies of the GCC. Findings reveals that ESD is often identified with environmental issues, thereby neglecting other SDGs and their core values. The solutions suggested in this chapter include further training and research on implementing SDGs and establishing a specific unit in each HE body to facilitate and supervise ESD implementation and SDGs integration into curricula.

Keywords

Education for sustainable development · Education policy · Higher education · Sustainable development

H. Mahfoodh (✉) · H. AlAtawi
University of Bahrain, Skhir, Bahrain
e-mail: hakadhem@uob.edu.bh

H. AlAtawi
e-mail: halatawi@uob.edu.bh

5.1 Introduction and Background

This chapter examines the factors leading to misinterpretations of Education for Sustainable Development (ESD) in the Cooperation Council for the Arab States of the Gulf (GCC) within the context of Higher Education (HE). ESD emerged as a result of the vibrant nature and fast-paced progress of the twenty-first century, with its ongoing and rapid technological changes, increasing environmental awareness, and a wide spectrum of innovation and technologies (Ramos et al. 2018). Simultaneously, human population growth, an increase in inequities, the fragile nature of economic systems and the accelerating problems of global warming have increased the problems and challenges continuously under investigation and global focus (Kabeer and Santos 2017). The increase in environmental awareness and sustainable development has not solved or minimised primary social, educational or environmental challenges. The reasons behind this failure vary, but one particular challenge is the concept of sustainability, a term framed according to national and international interests. This chapter explores the challenges faced when HE institutions and bodies adopt sustainability, especially within the GCC.

Sustainability and sustainable development are highly controversial in education and pedagogy. Whilst there is consensus that there is no single

K. Al'Abri and E. Buckner (eds.), *Higher Education and Sustainable Development in the Arab Gulf States*, Sustainable Development Goals Series, https://doi.org/10.1007/978-3-032-19559-3_5

definition for the term 'sustainable', HE bodies and institutes tend to link sustainable development to the development of society, the environment and the economy (United Nations 2002). The pedagogical and educational practices designed for sustainability and sustainable development need further articulation to be useful. According to the United Nations Educational, Scientific and Cultural Organisation (UNESCO), Education for Sustainable Development relies on the fundamental principles behind sustainable development, focusing on five key areas: environment, society, economy, peace and partnership (United Nations Educational 2016). ESD also requires applying an advanced level of integration and inclusion of the various issues and fields related to sustainable development at all levels of education. Common ESD topics include food and water security, global warming, resilience in times of crises, biodiversity, decreasing poverty and related issues. It can be said that ESD is capable of broad and various interpretations and formulations. Still, the term ESD can be a double-edged sword: it is powerful in its diverse articulations but also prone to being circumscribed to limited definitions.

Whilst ESD vows openness and equity, academic research shows that it needs further examination and analysis to achieve its basic promise: a better future for all. This chapter looks at the gap between an ideal ESD scenario and its current manifestation in the GCC states and suggests feasible recommendations that would lead to minimising this gap.

5.2 Literature Review

The term sustainable development emerged in academic and international environmental discussion as early as the 1980s amidst increased global awareness of the relationship between environmental issues and economic and social development (Agbedahin 2019; United Nations 1987a; United Nations 2002). Significantly, the Brundtland Report of the World Commission on Environment and Development created what was termed 'Our Common Future', adopted by the United Nations General Assembly in 1987, highlighting the importance of sustainable resources for future generations (McKeown et al. 2002; United Nations 1987b; United Nations Educational 2014). Despite the growth of its significance, sustainable development and sustainability lack a unified definition, but rather, many articulations (Agbedahin 2019, pp. 669–680). One common definition describes sustainable development as 'development that meets the needs of the present without compromising the ability of future generations to meet their own needs' (United Nations Educational 2014). It also 'requires meeting the basic needs of all and extending to all the opportunity to fulfil their aspirations for a better life' (United Nations Educational 2015). It is also associated with supporting the well-being of this green planet, Earth, and its people by creating reciprocal relationships between environment, society and economy, acknowledging the triple concerns of sustainable development: planet, prosperity and the people on the planet (McKeown 2002). More recently, sustainable development has been articulated beyond this classic triad, seeking to create partnerships amongst its various components, pursuing further equities so that all parties are winners (Sachs 2012; United Nations Educational 2015). This shift has led to the emergence of a new term, 'strong sustainability', relying on integrated and interconnected Sustainable Development Goals (SDGs), which aim to increase equity and inclusiveness amongst all constituents of the triad (Szyja 2018; Howe 2019). Although 'sustainable development' is a slippery term, it remains highly ambitious in that it aspires to sustain a world where 'no one is left behind' (United Nations Educational 2016). Consequently, this term has undergone ongoing shaping and reshaping, but has not achieved its promised equity and inclusiveness.

The concept of sustainable development has significantly influenced educational bodies. The term 'Education for Sustainable Development' (ESD) has emerged in the various Higher Education (HE) institutions and bodies adopting the SDGs. ESD is also associated with the seventeen SDGs of the UN, including equality, education, food security and other SDG goals. It also promotes teaching strategies that balance local

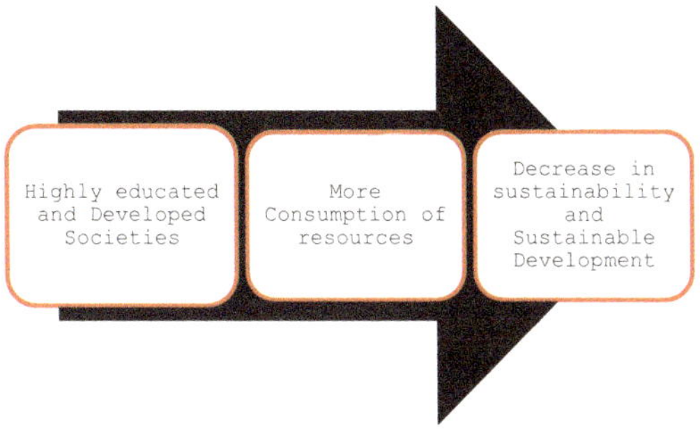

Fig. 5.1 Sustainable development process

priorities and global concerns, reflecting that ESD is substantially related to global requirements and conditions, but also accommodates local cultures and needs. Whilst many HE sectors and institutions express pride in their SDG endorsements, McKeown (2002) and McKeown and Hopkins (2003) argue that ESD is not enough to achieve sustainable societies because 'global consumption patterns show that the most educated societies leave the deepest ecological footprints' (p. 119). Consumption is a key factor in increasing unsustainable practices, which leads to the process shown in Fig. 5.1. In other words, highly educated and developed societies tend to have higher rates of consumption, which, in turn, leads to an inevitable decrease in sustainable resources and development.

In the GCC, ESD is closely associated with implementing SDGs, which also reveals the same challenges worldwide. Globally, SDGs are implemented using a 'cherry-pick' method, namely, according to irrational studies or insufficient factors in terms of prioritisation (Al-Saidi 2022). According to Heras-Saizarbitoria et al. (2022), uncalculated implementations of SDGs classically result in what is termed 'SDGwashing', meaning SDGs are considered accessories or shallow commitments rather than social, economic and environmental responsibilities. In addition, the GCC states experience inconsistencies when adopting SDGs in HE curricula and a lack of evidence-based framework when adopting ESD in the same context (Allen et al. 2017; Al-Saidi 2022). These problems are exacerbated by existing academic literature and field reports, which suffer from an absence of solid analysis of intersected assessment and observation (Bennich et al. 2020). Because of the increasing awareness of the significance of ESD in HE, various GCC states have experienced calls for immediate action to develop and prioritise SDGs on both local and regional levels. Some suggestions include developing peaceful cultures and societies and adopting feasible strategies to reduce inequalities, which Bissat and Rihan (2019) consider the path to sustainability in the GCC. Other suggestions encourage prioritising SDGs according to selective norms, including well-being, natural resources preservation, peace, equality and preserving the individual's dignity through decent living and careers (Allen et al. 2017). Other approaches underscore interrelated SDGs as a higher priority than others. For instance, Yang et al. (2020) argue that SDGs 1 (No Poverty), 2 (Zero Hunger) and 6 (Clean Water and Sanitation) are crucial to the sustainability of main supplies and the economic sustainability of societies and nations. The GCC states' inability to produce clear procedures or frameworks to prioritise SDG implementation leads to vague articulations of ESD, which also agrees with the global trends.

It is worth mentioning that the GCC states have adapted to globalisation, including the interest in sustainability. Issues such as innovative technology, international investment, free trade and modern waves of social change and development have become paramount in national strategies and policies associated with sustainable development (Fattouh and Sen 2021). These interests are traceable in the national visions of the respective GCC states, which are based on sustainable development and aligned with the SDGs; this interest has also created various national visions, including Bahrain Vision 2030, Kuwait Vision 2035 and Saudi Vision 2030 (International Monetary Fund 2023). Implementing these national visions entails a greater focus on education for sustainable development to prepare graduates to enter a marketplace oriented towards diversification and a knowledge-based economy supported by reliable and sustainable fundamentals and institutions (Ben Hassen 2022). The GCC states have adopted ESD, vowing to promote lifelong learning opportunities by considering cultural sensitivity, tangible content and knowledge related to the respected localities. They also emphasise the key strategies ESD promotes, including critical thinking, envisioning future scenarios, collaborative decision-making projects, supporting human rights, gender equity and fighting poverty (Bourn 2014). These strategies have significantly become a major focus in most HE sectors and organisations, often presented as goals within the national visions of the respective GCC states.

ESD requires further academic and pedagogical assessment and questioning to achieve effective integration in HE bodies in the GCC. Research indicates that since ESD is a broad term, it lacks clarity, leading to 'inefficiencies in achieving goals and development of initiatives' (Pavlova 2011, p. 333). Instead, it is often confused with Environmental Education (EE), which focuses on environmental issues exclusively (Menton et al. 2020). However, after the World Education Forum held at the UN in 2015, the concept of ESD changed dramatically through the Incheon Declaration, or the Education 2030 Agenda, emphasising the quality and relevance of education as a right for all people and the public good. This agenda supports an educational shift that puts UN SDGs at the centre of the educational process (United Nations 2015). The promises of ESD are fragile in a world that lacks security and equity due to economic, political and biological factors, and the GCC is no exception. The COVID-19 pandemic, for instance, led to various inequities in education, especially in regions labelled as underdeveloped or developing (Mahfoodh and AlAtawi 2021). In addition, implementing ESD is often subjected to personal interpretation, leading educators to convey content inadequately, another indication that ESD needs further reflection and assessment in education.

Whilst much research is conducted on sustainable economies in the region, more could be done in the regions' of HE. Lessons learned from online education during the pandemic can provide a starting point for developing sustainability in the HE curriculum, beginning with rethinking the different articulations of sustainability and education for sustainable development (Mahfoodh and AlAtawi 2020). Like the global experience of improving and assessing ESD in HE, the GCC states can benefit from previous experience and adopt successful strategies. However, this stage cannot be initiated without exploring recent manifestations of ESD and integrating SDGs in HE in the region. This chapter examines the factors leading to misinterpretations of ESD reflected in the GCC compared globally. It also examines the fetishising of SDGS and ESD in the GCC culture of HE institutions and bodies.

5.3 Research Objectives and Methods

The chapter adopts a GAP analysis, comparing the ideal ESD scenario and current practices that promote ESD in the HE sectors to address the existing gap in implementing ESD in the GCC. Ideal scenarios are surveyed through a literature review, whilst current ESD practices are based on both a literature review and a sample of 40 participants from the six GCC countries and international participants from the UK, the US and Germany. The participants are all educators in the HE sector, and they have all endorsed ESD in their courses. Twenty participants specialise in SDGs and Sustainability through partnerships with the UN programmes, whilst the other twenty are general educators who have adopted SDGs. Thirty participants are from the GCC, four from the UK, with the US and Germany represented by three participants each.

The main question of the chapter is: How can we bridge the gap between an ideal ESD scenario and its current manifestation in the GCC states through feasible and concise steps? In other words, how can ESD become an active tool that promotes diverse cultures and different forms of equity beyond the current SDGs models of environmental solutions? In short, this chapter addresses the challenges and common misunderstandings of ESD and concludes with some recommendations that promote and assess ESD practices in HE institutes and bodies.

Forty participants were selected to represent private and public HE bodies in the GCC and from the UK, the US and Germany. This selection aimed to provide a balanced perspective that includes both the private and public sectors and the GCC, compared to the international HE perspective. The selected participants are familiar with sustainability, and their affiliations have adopted frameworks and strategies to implement ESD and SDGs. In addition, the participants of this research willingly chose to participate because they expressed interest in sustainability and ESD. A drawback of the sample is the limited number of international experts available due to the time constraints of the research.

The data collected from respondents underwent two rounds of a specifically designed Delphi Protocol, which is composed of only two rounds with two panels: twenty experts and twenty practitioners of SDGs. In both rounds, each panel answers a set of questions, followed by a short report by the researchers. The researchers then share the report with both panels without mentioning names, and after the panels read the report, they answer the same questions in the second round. This methodology drives participants to become more self-critical and allows space for reflecting on their rationale. The main target is to achieve consensus or increase agreement in their responses. The participants were divided into two equal panels: Panel One, specialising in SDGs and sustainability, and Panel Two, representing educators who do not specialise in SDGs but still adopted ESD in their courses. After the first round, a short report was developed and cross-shared for reflection. The second round involved answering the same questions and providing space for personal reflection on ESD in HE curricula and pedagogical practices. This Delphi Protocol was followed by a group discussion categorised into Group 1, the GCC, and Group 2, the global participants. Each group had to reflect on possible weaknesses and opportunities to improve ESD in HE curricula and educational strategies. In addition, three senior managers affiliated with UNESCO's regional office were interviewed for further reflections on sustainability and ESD (Questions in the Appendix).

The participants' security was of great concern in this research through several protocols. First, all personal information was kept anonymous, without any reference to the name, rank and affiliation of all respondents. In addition, the Delphi Protocol was done individually, and the discussion group was on the Zoom platform. All participants willingly agreed to sign a consent form and a statement of any conflict of interest before answering the questionnaires in the Delphi Protocol and before the interviews.

5.4 Findings

This section adopts an integrated approach using the GAP model to explore the gap between the ideal scenario for implementing ESD effectively and current practices that claim to adopt ESD in the various HE sectors in the GCC. The Gap model is divided into three sections: current scenario, ideal scenario and bridging the GAP (in the Sect. 5.5). Integrating SDGs requires acknowledging the essential principles of HE institutions as contributors to 'the intellectual and moral improvement of the human condition' (Gough and Scott 2008). In addition, a thorough understanding of these principles allows HE to recognise its dominant role in spreading awareness and encouraging learners to be knowledgeable and creative whilst dealing with issues related to sustainable development, rather than urging them to take action to maintain sustainable development. Therefore, HE is responsible for providing learners with the fundamental knowledge of SDGs alongside twenty-first century skills and involves parallel ESD training for educators, facilitators and other members.

Since its emergence, ESD has promoted knowledge, skills and values that equip learners with the required knowledge and efficacy to make sound decisions and take action for a better world, characterised by commitments to a reliable environment, a powerful economy and a society based on justice and equity (United Nations Educational 2018). This idea means that HE should be closely integrated into sustainable development with specific attention to the 17 SDGs, entailing striking changes in the concept of education itself, such as abandoning the 'classic one-way delivery of knowledge by a teacher or lecturer through a written textbook in a classroom confirmed by written tests' (Agbedahin 2019). Instead, it encourages integrating various modes of education, including formal and nonformal, simulations and modelling and real-life scenarios and role-play experience, which amalgamate the essential competencies for sustainable development, such as peer review, teamwork and problem-solving (Mahfoodh and AlAtawi 2021; Mahfoodh and AlMuqahwi 2023). An ideal ESD in HE involves changing the culture and content of course delivery, assessment and curricula.

The interviewed experts of UNESCO in the GCC showed consensus on the need to reflect on the region's experiments of ESD implementation because it is an ongoing process subject to the national vision of each GCC state, which creates differing priorities and visions in the GCC. One interviewee explained that some ESD attempts exist in the region, yet are 'undercooked' because they are limited to online learning and constantly changing assessment systems. According to the same interviewee, what is missing to achieve an ideal implementation of ESD is students' understanding of sustainability values, including equal opportunity, zero poverty and integrating SDGs. Another expert added that HE students occasionally participate in competitions for innovation, such as Injaz in Bahrain and Oman, which encourages student role-plays as simulations of the marketplace, but these are part of their extra work during the course. This expert insisted that these participation opportunities should ideally be a permanent, rather than occasional, part of HE curricula. An ideal ESD entails close integration between SDGs and curricula as a necessity, not a bonus or extra task. One expert argued that various GCC universities could create courses as college requirements, focusing on sustainability by following similar experiences in international universities, examining courses on sustainability topics, from global warming and resilience to education and social responsibility. Another expert suggested that the GCC must work closely with UNESCO programmes to design HE curricula related to ESD by creating a unit in every HE body to focus on ESD projects and implementation. This unit would be responsible for creating ESD policies, regulations, projects, curricula and events.

An ideal ESD would also mean that curriculum and pedagogy are interconnected with learners' daily practices and experiences (Cotton et al. 2009). Achieving an ideal state of ESD in HE is conditional on significant factors, including economic security and resilience, access to a reliable resource with varying group sizes and a close relationship with the national and international marketplace (Mahfoodh 2021). After all, sustainability is attached to both national and international markets. Hence, in its ideal scenario, ESD should interconnect with all these components to prepare students for a sustainable future as a cultural lifestyle with integrated value in the marketplace.

Implementing ESD effectively involves integrating all 17 components of the United Nations' SDGs in HE. Research indicates an urgent need for more discussions and tailor-made strategies to situate ESD by implementing all SDGs (Cotton et al. 2009). However, according to Mahfoodh (2021), the main concern is that most educators are not familiar with the concept of 'change' in terms of ESD. This research surveyed a set of questions in a Delphi Protocol of two rounds to validate this claim. The two panels in the first round were in agreement with the claim of 'lack of awareness regarding ESD' because twenty-eight participants articulated ESD in the context of 'going green' by encouraging students to make planting projects on special occasions or cleaning some public areas as part of environmental responsibility (See Fig. 5.2). Although this response reflects an understanding of the close relationship with a sustainable natural environment, it lacks the other major components of sustainability: economics and equities. Eight participants highlighted the problematic definition of ESD: how this term can be differently defined, depending on the national and international requirements and needs. Six other participants indicated that ESD is often confused with Environmental Education (EE), which again

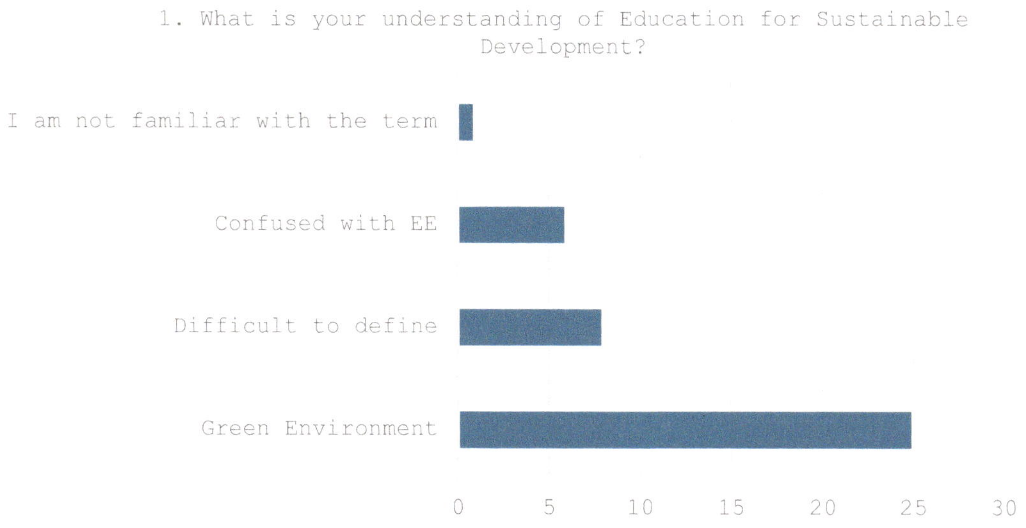

Fig. 5.2 Delphi protocol, round one, both panels—question 1

reveals misunderstandings of the main principles of ESD: equity and inclusion.

The two panels agreed on the students' response to SDGs when integrated into HE courses: students lack the required knowledge and culture of sustainability and sustainable development, as expressed in their notes. Thirty-two participants acknowledged that students still need more awareness of SDGs and sustainability (Fig. 5.3). They also reflected that students show interest when SDGs are linked to the marketplace and job placement, even when they do not fully understand such terminology. Some expert panel educators noted that the assessment of these courses needs to be modernised so that students grow to favour these courses, suggesting that grading and assessment need to go beyond the traditional marking of grammar, language and other formatting mistakes. Instead, the expert panel suggested an assessment based on effective solutions for real-world problems and feasible solutions that align with the respective state's national vision.

However, the two panels had varying responses to the challenges they faced in courses integrating SDGs. The expert panel considered students the

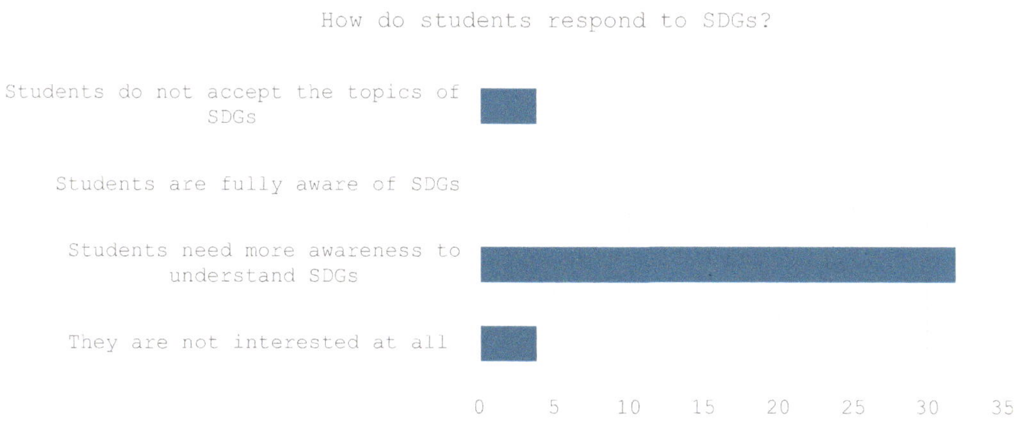

Fig. 5.3 Delphi protocol, round one, both panels—question 2

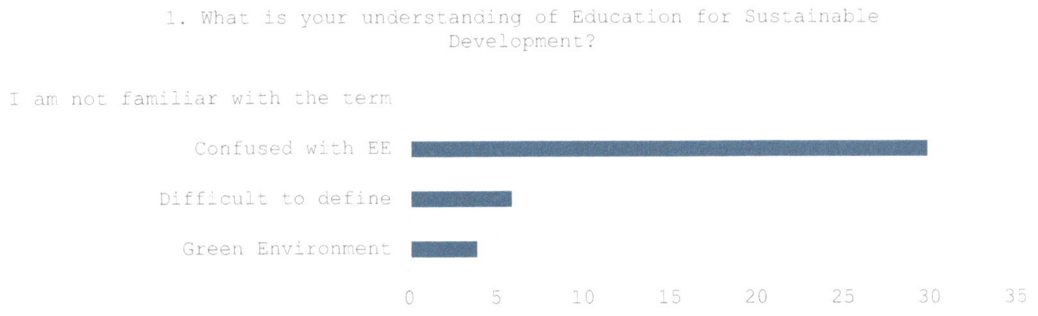

Fig. 5.4 Delphi Protocol, Round One, Both Panels—Q1

major challenge when implementing SDGs in the HE curricula (Fig. 5.4). They also noted that students find these courses inconvenient and sometimes uncomfortable because they are not used to the unconventional assessment and course delivery methods that ESD entails. In addition, five participants from the expert panel explained that students find it difficult to brainstorm and explore innovative and novel world-problem solutions since they are more used to practice-based knowledge and rehearsals for assessment. On the other hand, the second panel showed a lack of agreement, with the majority of eight educators considering the content as the main challenge in ESD; three educators noted that ESD content is difficult to adapt or integrate into some fields, including mathematics, law and Islamic studies, insisting that the students cannot see how SDGs can be linked to their core courses. These notes reflect a lack of SDG knowledge, confirming that the students are the main challenge when implementing ESD.

There is a general awareness of the significance of improving ESD in the GCC, which aligns with the literature review. According to the feedback of both panels, whilst some countries, including Oman and Bahrain, adopt conferences and selective events as a form of implementing ESD in their universities, other countries have created departments for environmental studies, such as some universities in KSA and Qatar. However, the expert panel argued that the HE bodies need to implement ESD as an ongoing part of developing curricula. They also explained that ESD is subjected to assessment and change to meet market needs and national visions. Panel Two suggested more conferences and seminars, workshops, collaboration with UNESCO and field trips to real firms in the marketplace.

In Round Two, the responses to questions Three and Four showed little change (see Appendix), but the answer to the first question witnessed a dramatic change (Fig. 5.4). Thirty educators reflected that ESD is often confused with Environmental Education (EE), especially when ESD is circumscribed exclusively to topics related to the green environment. The expert panel members explained that topics related to the natural environment dominate the literature on ESD, which makes most educators focus on these issues, at the expense of other SDGs, especially Zero Poverty and Equality. However, the interviewees from UNESCO's regional office explained that this issue could be solved through further training on SDGs, with a focus on the SDGs targeted by the various GCC national visions.

Further, the participants of both panels changed their responses to an almost 90% consensus that the students comprise the main challenge when implementing ESD in HE curricula, as shown in Fig. 5.5. Panel One repeated the same notes in Round One and added that most students are used to traditional quizzes and exams, whilst ESD assessment is more project-based and involves teamwork. They also added that students lack direct contact with the real-world marketplace in the GCC states, unlike the US, the UK and Germany, where most students study and work at the same time. This difference in students' career experience distances them from mastering

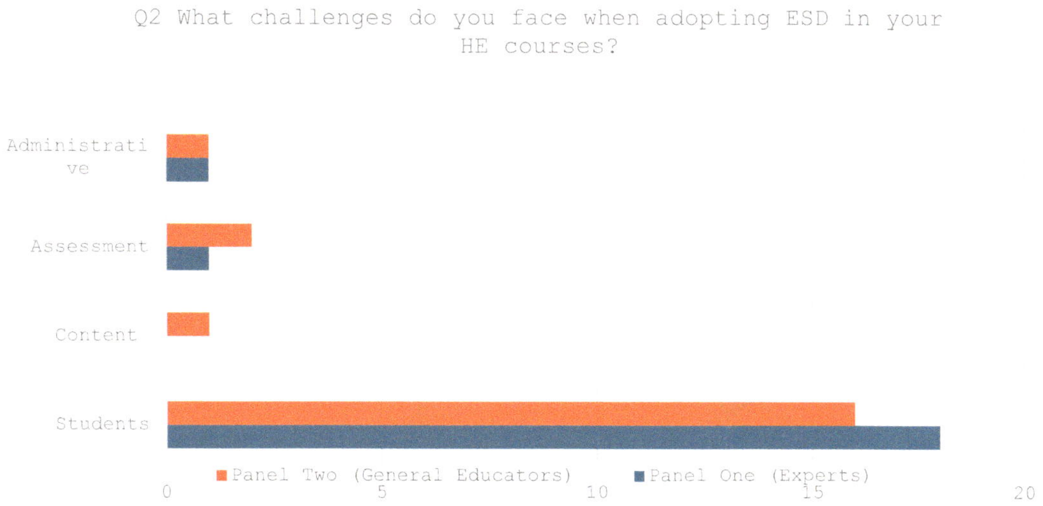

Fig. 5.5 Delphi protocol, round two, both panels—Q2

the skills required for real-world problems and the practical life they experience in the labour market. In addition, participants had the chance to reflect on the flaws they experienced or noticed when implementing ESD. Some educators complained that SDGs are delivered as volunteering activities for which students receive bonuses. Activities include occasional trips to clean beaches, planting flowers for public occasions and casual visits to clean a certain venue. Other activities include participating in regional and local environmental contests related to renewable energy and environmental issues. Participants agreed that temporary, occasional activities that are limited to certain students contradict the idea of sustainability. Such practices create a culture that misunderstands sustainable development as an ongoing process for which all people are responsible, thereby turning ESD into a fashionable lifestyle or a commodity used temporarily and then neglected. Hence, these activities lead to contradictory results. For instance, when volunteers grow seasonal plants, they usually use plastic cups and pots, which 'is only another form of increasing pollution. It is like you kill your own environment', according to one of the participants. Another participant explained that the idea of volunteers and choosing participants for the contest is selective, eventually breaching SDG 10 (Reduced Inequalities). In short, implementing ESD has turned SDGs into commodities since they are temporary and consumed rather than sustainable and constantly developing.

Some HE educators reflected that many ESD practitioners consider online learning the only means of integrating SDGs into curricula. This misunderstanding led to focusing on course design and allocation of marks rather than a real integration of ESD. One participant insisted that COVID-19 has led to articulating ESD within the limits of online education, which is sustainable to some extent but also excludes the seventeen SDGs.

These reflections were categorised into two overarching categories, namely challenges for educators and challenges for students. Challenges for educators included issues with: defining the SDGs, prioritising SDGs, designing assessment for students, organisational and administrative problems and their own SDG and ESD literacy. Meanwhile, challenges for students included difficulties doing project-based tasks based on real-world problem-solving, a lack of real work experience, underdeveloped communication skills and a lack of critical thinking skills.

5.5 Discussion and Analysis

Following the Delphi Protocol, all participants had a virtual discussion facilitated by the researchers. The discussion examined suggestions to bridge the gap between the two scenarios and lead towards better practices in implementing ESD and SDGs. The UK and US participants suggested that training and education are common ways to bridge this gap. They added that the training should be tailor-made for students, focusing on ways to introduce HE students to the real-world marketplace. However, the participants from Germany extended the urgent need for further training to both HE educators and students. They also suggested further partnerships with UNESCO and international universities implementing ESD and integrated SDGs. The GCC participants stressed that training needs to be supported by a genuine will to integrate SDGs in all disciplines as an integral part of the assessment and curricula beyond occasional events. These steps need to be part of the HE policies and regulations, designed precisely to support the respective national visions. Twelve participants suggested that ESD should be part of the ongoing CPD for HE faculty to avoid confusion amongst students and educators, and that ESD should be linked to the annual appraisals of HE educators. Further, thirteen participants insisted that there should be a semi-annual or annual assessment of both the faculty and managerial levels of HE. By doing so, ESD can be improved based on the assessment and the changing marketplace needs.

From the participants' feedback, it can be inferred that ESD needs to replace 'cherry-picked' procedures with ongoing and permanent protocols and training, considering both the individual and the professional. The whole concept of ESD should go beyond the occasional conferences or workshops conducted to improve university rankings or to fill in the schedules of students' activities. The expert panel explained that ESD means 'to be sustainable, lasting, durable and surviving, which means that SDGs need to be felt, experienced and provoked within the classroom and amongst HE educators alike'. Furthering this idea, one of UNESCO's representatives argued that SDGs should be seriously integrated into curricula after a careful assessment that adapts to the national vision of each GCC state and argued that 'most HE bodies had some initiatives in ESD, but they do not follow clear objectives or a unified national vision. Instead, these initiatives are more individual efforts of HE educators interested in climate change and renewable energy issues in the GCC'. One of the experts explained that the University of Bahrain plans to create a course that involves ESD with close mentoring from the local UN office. Another expert explained that some postgraduate sustainability programmes were developed recently in the region and may need further assessment and more development to meet market needs, drawing on their experience in Saudi Arabia.

The findings also suggest a lack of a unified framework or policy on implementing ESD and effectively adopting SDGs in HE in the GCC states. To solve this problem, the researchers of this paper suggested creating a specific unit or office in each university that fosters and monitors the development of ESD. Such an office could include the mandate of ESD and Environmental Research, focusing on renewable energy along with the seventeen SDGs. Priority should be given to the SDGs directly related to the national vision of each state. However, this office should create strategies and regulations that assure an effective integration of SDGs, which means allocating a certain percentage of coursework/ assessment to project-based work where students need to solve a real-world problem related to these SDGs. Still, it is worth mentioning that SDGs can be adopted in all fields because they range from the scientific, such as Clean Water and Sanitation (SDG 6), to the environmental and social, including Reduced Inequality (SDG 10). What the HE needs is further articulation and discussion between their various faculty members and the management to explore alternative articulations of SDGs and potential methods of integrating them into all HE disciplines.

In addition to the above-suggested unit, some steps can be adapted to turn implementing SDGs into the HE curriculum into a successful learning experience for both educators and learners. First and foremost, HE needs to collaborate with other governmental institutions and agencies to establish a generic framework that will reflect the government's vision without endangering HE roles and principles. These collaborations should provide paid internships and training opportunities for educators and students, which will help bridge the gap between students and the marketplace. A further step into providing Continuous Professional Development (CPD) and training to faculty members, curriculum developers and educators is required to resolve any confusion and supply them with ongoing updates in knowledge and technology. This step needs to be part of the compulsory CPD of all academic faculty members rather than a choice. Finally, it is wise to include the academic faculty in the national and international policies regulating and administering ESD because the educator is the basis of ESD delivery and success.

5.6 Conclusions and Implications

This chapter explored the current situation of ESD in the various HE bodies of the GCC. According to the findings, ESD is misunderstood by exclusively circumscribing sustainability to environmental issues and green energy. In addition, it was found that ESD is implemented in the form of casual conferences and forums rather than articulated planning, and the curricula of HE lack systematic approaches to adopting ESD. After processing the gathered data, this research suggests a significant need to revisit various articulations of ESD and its situation in HE today. According to the above discussion, there is a genuine need to explore and reconsider how SDGs are delivered, integrated and practised in the various HE curricula as part of implementing ESD. Because these terms are relatively new to academia and pedagogy, the academic faculty members and HE administrations can play a more positive and vital role in disseminating this knowledge, especially since educational sectors are usually closely attached to the marketplace in their visions and strategies. Still, this process needs tremendous training and more engagement with events and conferences on ESD and sustainability topics.

We suggest establishing a specific unit in every university responsible for implementing ESD and SDGs effectively. This unit acts as a facilitator that bridges the gap between HE and the marketplace, creating room to engage HE educators in constructing and contextualising national and international policies and discourse based on ESD practices and strategies. The academic faculty members are capable of adopting and establishing effective visions and scopes to articulate successful definitions of ESD by adapting their critical thinking skills and methods, but their efforts and practice need to be organised and collective-based to support the national vision of each state. In addition, since this unit is directly connected with the marketplace, it should involve students as the focal point of their initiatives. In this way, students will gain a deeper understanding of SDGs through real-life situations vis-à-vis the marketplace. Consequently, with its efforts and activities, this unit could lead to a better understanding and implementation of sustainable development, thereby creating a more articulate and solid ESD environment and practice, serving all related stakeholders: students, faculty, the University and the official bodies of the marketplace.

It should be noted that it is necessary to evaluate the current HE courses to ensure the integration of all SDGs into various curricula and to implement skills for the twenty-first century. Further, HE's cooperation with governmental offices, private sector firms, international SDG bodies and organisations and society would also improve the situation of ESD in the GCC. Additional drawbacks could be solved through further training, research and active communication between the various components involved in HE practices of ESD and SDGs implementation.

References

Agbedahin AV (2019) Sustainable development, education for sustainable development, and the 2030 agenda for sustainable development: emergence, efficacy, eminence, and future. Sustain Develop 27(4):669–680. https://doi.org/10.1002/sd.1931

Allen C, Nejdawi R, El-Baba J, Hamati K, Metternicht G, Wiedmann T (2017) Indicator-based assessments of progress towards the sustainable development goals (SDGs): a case study from the Arab region. Sustain Sci 12:975–989. https://doi.org/10.1007/s11625-017-0437-1

Al-Saidi M (2022) Disentangling the SDGs agenda in the GCC region: priority targets and core areas for environmental action. Front Environ Sci 10:1025337. https://doi.org/10.3389/fenvs.2022.1025337

Ben Hassen T (2022) The GCC economies in the wake of COVID-19: toward post-oil sustainable knowledge-based economies? Sustainability 14(18):11251. https://doi.org/10.3390/su141811251

Bennich T, Weitz N, Carlsen H (2020) Deciphering the scientific literature on SDG interactions: a review and reading guide. Sci Total Environ 728:138405. https://doi.org/10.1016/j.scitotenv.2020.138405

Bissat LM, Rihan C (2019) Implementing agenda 2030 in the Arab world: contextualisation, action areas and policy planning. Public Sector Econ 43(4):459–476. https://doi.org/10.3326/pse.43.4.7

Bourn D (2014) The theory and practice of development education: a pedagogy for global social justice. Routledge, New York

Cotton D, Bailey I, Warren M, Bissell S (2009) Revolutions and second-best solutions: education for sustainable development in higher education. Stud High Educ 34(7):719–733. https://doi.org/10.1080/03075070802641552

Fattouh B, Sen A (2021) Economic diversification in Arab oil-exporting countries in the context of peak oil and the energy transition. In: Giacomo L, Moerenhout T (eds) When can oil economies be deemed sustainable? Palgrave Macmillan, Singapore, pp 73–97

Gough S, Scott W (2008) Higher education and sustainable development: paradox and possibility. Routledge, New York

Heras-Saizarbitoria I, Urbieta L, Boiral O (2022) Organisations' engagement with sustainable development goals: from cherry-picking to SDG-washing? Corp Soc Responsib Environ Manag 29(2):316–328

Howe P (2019) The triple nexus: a potential approach to supporting the achievement of the sustainable development goals? World Dev 124:104629. https://doi.org/10.1016/j.worlddev.2019.104629

International Monetary Fund (2023) Gulf cooperation council: economic prospects and policy challenges for the GCC countries. https://www.imf.org/en/Publications/CR/Issues/2023/12/14/Gulf-Cooperation-Council-Economic-Prospects-and-Policy-Challenges-for-the-GCC-Countries-542513

Kabeer N, Santos R (2017) Intersecting inequalities and the sustainable development goals: insights from Brazil. WIDER 12:1–47

Mahfoodh H, AlAtawi H (2020) Sustaining higher education through e-learning in post Covid-19. In: Proceedings of the 2020 sixth international conference on e-learning (econf). IEEE, pp 361–365. https://doi.org/10.1109/econf51404.2020.9385477

Mahfoodh H, AlAtawi H (2021) Higher education in COVID-19: from emergency to sustainable remote education. In: Proceedings of the 2021 sustainable leadership and academic excellence international conference (SLAE). IEEE, pp 1–7. https://doi.org/10.1109/slae54202.2021.9788076

Mahfoodh H (2021) Reflections on online EFL assessment: challenges and solutions. In: Proceedings of the 2021 sustainable leadership and academic excellence international conference (SLAE). IEEE, pp 1–8. https://doi.org/10.1109/slae54202.2021.9788097

Mahfoodh H, Almuqahwi RIS (2023) Potential reforms in the GCC higher education post COVID-19. In: Segumpan R, McAlaney J (eds) Challenges and reforms in gulf higher education: confronting the -COVID-19 pandemic and assessing future implications, Routledge, London, pp 7–25

McKeown R (2002) Progress has been made in education for sustainable development. Appl Environ Educ Commun Int J 1(1):21–23. https://doi.org/10.1080/15330150213983

McKeown R, Hopkins C (2003) EE p ESD: defusing the worry. Environ Educ Res 9(1):117–128. https://doi.org/10.1080/13504620303469

McKeown R, Hopkins CA, Rizi R, Chrystalbridge M (2002) Education for sustainable development toolkit. Knoxville: Energy, Environment and Resources Center, University of Tennessee

Menton M, Larrea C, Latorre S, Martinez-Alier J, Peck M, Temper L, Walter M (2020) Environmental justice and the SDGs: from synergies to gaps and contradictions. Sustain Sci 15:1621–1636. https://doi.org/10.1007/s11625-020-00789-8

Pavlova M (2011) Environmental education and/or education for sustainable development: what role for technology education? Perspect Learn des Technol Educ 25(8):333–339

Ramos TB, Caeiro S, Moreno Pires S, Videira N (2018) How are new sustainable development approaches responding to societal challenges? Sustain Dev 26(2):117–121. https://doi.org/10.1002/sd.1730

Sachs JD (2012) From millennium development goals to sustainable development goals. Lancet 379(9832):2206–2211. https://doi.org/10.1016/s0140-6736(12)60685-0

Szyja P (2018) The role of the concept of sustainable development and its indicators in a study on the progress of the new UN goals. Econ Environ Stud 18(45):387–409

United Nations Educational, Scientific and Cultural Organization (2014) Shaping the future we want: UN decade of education for sustainable development (2005–2014) final report. UNESCO, Paris. https://unesdoc.unesco.org/ark:/48223/pf0000230171

United Nations Educational, Scientific and Cultural Organization (2015) Rethinking education: towards a global common good? UNESCO, Paris. https://unesdoc.unesco.org/ark:/48223/pf0000232555

United Nations Educational, Scientific and Cultural Organization (2016) Global education monitoring report-place: inclusive and sustainable cities. UNESCO, Paris. https://unesdoc.unesco.org/ark:/48223/pf0000246230

United Nations Educational, Scientific and Cultural Organization (2018) Issues and trends in education for sustainable development. UNESCO, Paris. https://en.unesco.org/sites/default/files/issues_0.pdf

United Nations (1987a) Report of the world commission on environment and development note: by the secretary-general. https://digitallibrary.un.org/record/139811?ln=en

United Nations (1987b) Report of the world commission on environment and development: our common future. www.un-documents.net/our-common-future.pdf:UnitedNations-WCED-WorldCommissiononEnvironmentandDevelopment

United Nations (2002) Report of the world summit on sustainable development. In: World summit on sustainable development. United Nations, Johannesburg, pp 1–174. https://digitallibrary.un.org/record/478154?ln=en

Yang S, Zhao W, Liu Y, Cherubini F, Fu B, Pereira P (2020) Prioritising sustainable development goals and linking them to ecosystem services: a global expert's knowledge evaluation. Geogr Sustain 1(4):321–330. https://doi.org/10.1016/j.geosus.2020.09.004

Hajar Mahfoodh Assistant Professor, obtained her PhD in English Literature from the University of Surrey and teaches English at the University of Bahrain. She is specifically keen on research exploring colonialism and decolonising curricula, postcolonial culture and higher education, sustainable development in higher education in MENA, and developing alternative policies and frameworks that could effectively reform higher education in the GCC region.

Hessa AlAtawi is an English lecturer at the English Language Centre, University of Bahrain. She received her MA degree in American Studies from the University of Jordan (2004). Her academic interests include English, eLearning, and Higher education reform, and she has published many papers on various topics, including education, culture, and sustainable development.

6 Faculty Members' Perceptions of Higher Education for Sustainable Development: A Study from the College of Education at Qatar University

Mayamin Altae

Abstract

Sustainable Development (SD) has become a focus for higher education over the past two decades, with institutions worldwide seeking new approaches to advance sustainability goals. Like other nations, GCC countries have developed national visions aligned with the United Nations' 2030 Agenda. In Qatar, the Qatar National Vision 2030 emphasises sustainability across all sectors, including education. As the country's main teacher-training institution, Qatar University's College of Education plays a vital role in promoting understanding of SD amongst educators. This study explores faculty perceptions of Education for Sustainable Development (ESD) through semi-structured interviews with two faculty members responsible for teacher preparation. The interviews examined participants' understanding of ESD and their ability to integrate sustainability and climate change concepts into teaching materials. Findings reveal that faculty members demonstrate limited awareness of how to incorporate climate change issues into their courses and undervalue the potential of these topics to inspire students as agents of change. The study concludes by offering recommendations to strengthen faculty capacity, enhance their comprehension of sustainability and better equip them to embed ESD principles into teacher education programmes.

Keywords

Sustainable development · Higher education · Environmental awareness · Student teachers

6.1 Introduction

The emergence of technology as a dynamic global force in the recent decades has led to an increase in competition amongst worldwide markets to obtain larger shares of resources to support their operational activities across the globe. Little attention has been given to the pressure on resources that this competition mania has caused due to focusing on making more profit without considering the harmful impact on natural resources, poverty, pollution and inequality. Countries across the world have had their negative contributions to this environmental disaster in their attempts to draw more resources to support their national growth. The world has recently recognised the self-destruction approach that prevailed over the last few decades and a new approach has started to evolve in the midst of the current environmental problems, an approach of sustain-

M. Altae (✉)
Faculty of Education, The University of Buckingham, Buckingham, UK
e-mail: mayamin.altae@buckingham.ac.uk

The University of Buckingham, Buckingham, UK

K. Al'Abri and E. Buckner (eds.), *Higher Education and Sustainable Development in the Arab Gulf States*, Sustainable Development Goals Series, https://doi.org/10.1007/978-3-032-19559-3_6

able development to reverse the destruction attitude and slow it down. The traditional growth practices strive to achieve growth at any cost whilst the new approach gives more attention to social justice and environmental development. The higher education sector, represented by universities, has emerged as a potential platform to promote sustainable development in communities due to its direct impact on raising awareness and change outdated habits. In that, the colleges of educations are seen as effective platforms for change due to their direct involvement in training teachers who have more influence on communities through raising a generation of children with capabilities to understand the importance of sustainable development and who can drive the paradigm of change in this respect. More specifically the context of this paper is the role of the College of Education at Qatar University. The college has been advocating for sustainability concerning environmental development, equal opportunities for all and more recently being the pioneer in promoting the wellbeing aspects as a roadmap to support environmentally friendly initiatives and educate future generations about the importance of applying new methods to support the country growth in their workplaces. This also addresses the fundamentals of Qatar National Vision (QNV) 2030 that aims at directing Qatar 'towards a balance between developmental needs and the protection of its natural environment, whether land, sea or air' through creating a framework and institutions 'that can serve as the guardians of Qatar's environmental heritage. The QNV 2030 also emphasises the importance of increasing citizens' awareness of their role in protecting the country's environment for their children and the nation's future generations' (Planning & Statistics Authority). Thus, this paper attempts to examine the role of the College of Education at Qatar University (represented by the author being an academic at the College) in adapting its teaching practices to comply with a future education for sustainable development, particularly its educational approaches and main competencies needed for cleaner and sustainable growth. This is done through exploring the College faculty's level of understanding of sustainability, considering whether their knowledge is fragmented across separate concepts or reflects a comprehensive, whole-picture perspective. The paper aims to suggest some recommendations for making sustainable development an integral part of the college's performance as an educational institution.

There is a need for the College to robustly move towards reviewing the current programmes and update the list of courses offered to student teachers. Programmes that focus on global citizenship and sustainable education have the potential to change the scene of teacher-training policy in the country and contribute to the national vision of the country. Promoting innovative educational approaches, such as the transformational education approach, can support Sustainable Development practices; however, this needs to be monitored to measure influence by setting up certain performance indicators. For the scope of this paper and to narrow down the broader concept of sustainability, the author focuses, through self and chats with academic colleagues, on exploring the understanding of various statements on sustainable development (climate, economic aspects, social factors) to find out whether there is a separated understanding of the different components of sustainability or a whole picture understanding.

Sustainability has been interpreted in many different ways, especially in the field of education for sustainable development. The UN sustainable development goal 4 in article 7 states;

'By 2030, ensure that all learners acquire the knowledge and skills needed to promote sustainable development, including, amongst others, through education for sustainable development and sustainable lifestyles, human rights, gender equality, promotion of a culture of peace and non-violence, global citizenship and appreciation of cultural diversity and of culture's contribution to sustainable development' (UN). The UN SDG 4.7 also highlights the main components of sustainable development by addressing the issue of Global Citizenship in education, stating that:

> The role of Global Citizenship Education (GCED) is very relevant to the realisation of the 2030 Agenda for Sustainable Development. The three areas of sustainability, namely economic growth,

social inclusion and environmental protection, rely on the contribution of informed citizens (https://ocm.iccrom.org/)

Education for sustainability puts the three areas of social inclusion, economic growth and environmental protection at the heart of its focus. Education for Sustainability is considered, in various systems 'an integrative concept of environmental education where the ecological dimension of environmental education is more strongly supplemented with social and economic dimensions' (Para et al. 2020, p. 150).

6.2 Literature Review

The term 'sustainable development' has been discussed broadly in research. It also seems that different countries across the globe have started to focus on sustainable development differently. Therefore, understand that the aims of sustainable development may vary globally, and there may be a need to focus on the specific context where it is primarily promoted in the educational field. The Sustainable Development Goals were introduced by the United Nations back in 2015 as a global urge for countries to take actions 'to end poverty, protect the planet and ensure that by 2030 all people enjoy peace and prosperity'. The UN has also highlighted that 'the creativity, know-how, technology and financial resources from all of society are necessary to achieve the SDGs in every context'.

Sustainable development emerged as a concept due to environmental and ecological crises, and it has become the driving force to raise awareness on these issues since the beginning of the twentieth century (Du Pisani 2006). The increasing demands on exhausting the environmental resources towards the end of the twentieth century and the beginning of the twenty-first century raised the alarm that measures have to be taken to stop the deterioration of the ecological system. Such demands culminated in the discovery of oil, mainly when oil became the main force of industrial mobilisation activities and the main energy driver across the globe. This surge in oil consumption triggered another alarm about the exhaustion of such a vital source of energy. Hence, there was an increased demand to rethink the arbitrary usage of energy sources, and there were many emerging calls to adopt a more sustainable approach to the use of environmental sources. The limitations of the raw materials became a warning sign 'against wasteful consumptions' (Van Zon 2002: 110), hence 'sustainable development' emerged as a term to denote the chaotic waste of raw natural resources (Veblen 1917). Although an agreed-upon definition of 'sustainable development' has not been achieved yet, most of the interpretations tend to refer to 'sustainable development' as 'a pathway to all that is good and desirable in society' (Holden et al. 2014, p. 130). Such interpretation, however, is still confusing and broad, as it does not finely define what sustainable development really is. This complexity in finding a straightforward, clear and concise definition has led to confusion on the part of policymakers, who are struggling to exactly understand what policies are effective to address the current crisis. This has also resulted in some research raising the alarm that, without a clear and agreed-upon definition, sustainable development might end up being irrelevant (Redclift 2005).

Nevertheless, sustainable development remains the focus of the current policies for countries across the globe as the United Nations member states agreed to adopt it in 2012 at the United Nation Rio de Janeiro conference. Therefore, sustainable development remains a priority for the world, and most countries have set up their national vision to achieve that. In addition, the views on sustainable development have changed since the term was used at the end of the twentieth century. The understanding has shifted from being merely focused on ecological activities, such as recycling into more comprehensive in meeting the needs of the countries' development. Sustainable development in the early eighties, specifically in 1987, was looked at as 'development that meets the needs of the present without compromising the ability of future generations to meet their own needs' (Hajian and Kashani 2021: 2). The 1987 view reflected a more flexible and practicable approach to enable countries to meet their national visions.

That approach on sustainable development was further extended in 2019 to consider 'the needs of current and future generations, depending on appropriate circumstances of human, natural and economic capital to provide for human welfare' (ibid., 2). The new view takes into consideration the economic, the financial, the environmental and the social aspects as being important factors to consider when drafting policies to meet sustainable development goals.

It appears that university faculty members perceive sustainable development differently and according to their own and their institutions' viewpoints. Other factors influence the perceptions of faculty members such as media and work contexts. According to Samalisto and Lindhvist (2008), some faculty members change their teaching methods to address sustainable development requirements because their institutions have a high calibre about sustainable development, especially the environmental aspects, and such attitude can lead to instilling certain concepts in faculty members whilst delivering their teaching. Although the environmental aspects seem to be the prevailing ones amongst faculty members in promoting sustainable development practices (due to their plurality), other aspects can be integrated and introduced as being as important to achieve a comprehensive view. Faculty members can include other key aspects of sustainable development and present them to their students to raise awareness and meet the institutions' objectives. It is believed that faculty normally understand the various aspects of sustainable development in general; however, their perceptions about the relevance of such aspects to their teaching remit vary. Faculty must equip themselves with the required knowledge about sustainable development to be able to integrate it into their teaching practice and make it part of their subject materials. It does not have to be a component in the curriculum, or a standalone unit within the curriculum, rather sustainable development can be taught alongside the set curriculum by raising awareness about the issue as faculty go through their teaching norms. Faculty can disseminate sustainable development concepts to their students if they have enough knowledge about the topic, and colleges can facilitate this by developing a system within their teaching practices to raise awareness about the subject of sustainable development and integrate it into the routine teaching practices. Once the integration is achieved, a sustainable development intelligence database/system can be created where students can be tasked to think creatively about sustainable development topics and promote them externally.

6.3 Methodology

The population of the study are four faculty members who work at the College of Education, Qatar University. The participants meet the selection criteria as they train teachers, and they are involved in setting up programmes and courses related to teachers' training. Prior to choosing the four faculty members, I applied a random sample approach, which resulted in identifying the participants due to their close links to the training programmes and their involvement in modernising the curriculum. The four faculty members were good representation of the College staff.

As the aim of the study is to explore the faculty understanding of sustainable development and consequently integrate this into their teaching practice to influence their students to adopt a more sustainable approach in the future, the study uses a qualitative method through implementing a semi-structured interview as a tool. The four participants respond to some questions about sustainable development to express their views about the topic and their understanding. The responses are meant to explore the faculty's views about sustainable development and what they believe is considered part of the sustainable development concept. It was important to encourage the participants to think about sustainable development and express their understanding of the issue; therefore, the semi-structured interview was suitable as it allowed the researcher to gather more relevant data by asking additional follow-up questions. Galleta (2013: 24) mentions:

> Semi-structured interviews provide a repertoire of possibilities. It is sufficiently structured to address

> specific topics related to the phenomenon of study, whilst leaving space for participants to offer new meanings to the study focus. There is a great deal of versatility in the semi-structured interview and the arrangement of questions may be structured to yield considerable and often multi-dimensional streams of data.

Nonetheless, using the semi-structure interview as a tool has its limitations due to anonymity issues, i.e. participants might feel that their identity is revealed as the researcher conducted face-to-face interactions. This was addressed by providing some assurance about confidentiality through a detailed explanation of the process.

The responses were examined according to the participants' understanding of what is most and least relevant to sustainable development. In doing so, the faculty's level of understanding of sustainable development is revealed. More information is gathered about whether such understanding has directly affected their teaching methods and consequently influenced their students to follow suit. The questions that the participants responded to are:

- What is your understanding of climate change?
- What is your understanding of sustainable development?
- What is your understanding of education for sustainable development?
- How do you think your teaching materials cover issues related to climate change?
- How do you think your teaching materials cover topics related to sustainable development?
- What is your approach to integrating climate change and sustainable development aspects into your teaching materials?
- How do you think that integrating climate change and sustainable development aspects into your teaching would make a difference in your students' attitude to support such matters?

A thematic analysis followed the data collection phase with the primary focus on identifying specific patterns in the responses of the two participants through generating codes from data grouping. One of the main features of thematic analysis is 'the systematic process of coding, examining of meaning and provision of a description of the social reality through the creation of theme' (Vaismoradi et al. 2016, p. 100). However, I was cautious about the fact that I needed to be close to my data to be able to understand the direct and the indirect perceptions about sustainable development and climate change mentioned by the participants. Vaismoradi et al. (2016: 103) say that 'after achieving closeness to their data, researchers look for ideas in the data to recognise explicit and implicit ideas in the transcription based on their judgement'. Following an inductive thematic analysis that enabled identifying patterns (moving from particular to general) in the data, I formulated a sound understanding of the participants' responses that led to specifying three main themes. First, there is little understanding of how to integrate climate change and SD topics into the teaching materials. Second, there is little understanding about the importance of integrating climate change and SD topics into the teaching materials. Third, there is little understanding about enabling university students to be the catalysts of change to respond to future challenges related to climate change and SD issues.

6.4 Findings

The thematic analysis revealed few concerns and mixed understandings amongst the participants regarding addressing climate change and SD issues. There seems to be a prevailing perception that climate change and SD matters are 'nothing to do with me', i.e. the participants felt that it was not their responsibility to address such issues, as they only focused on their core teaching materials.

6.4.1 Climate Change and SD Matters Are not Being Integrated into Teaching

The data revealed that the participants do not integrate any issues about climate change and SD into their teaching. This is because they believe that such topics are irrelevant to their core subjects: 'No, I do not consider climate change matters relevant to my course ... I hear a lot about such issues, but how do I integrate them in my course?' (Participant 1). The participants believe that their focus needs to be on fulfilling the requirements of their courses and not to integrate other topics, which they believed to be irrelevant. They also expressed their concerns about integrating climate change and SD matters into their teaching because they have not been trained to do so, and they do not know how to address this:

> There is a lot of talks about climate change and SD matters everywhere; however, this has not been addressed officially with staff … If you ask me to integrate climate change and SD topics into my teaching, I simply do not know how to do this … they [climate change and SD matters] need to be presented to us officially with some training on how to achieve the integration (Participant 2)

The above response from participant 2 was concomitant with another participant's response, who clearly stated:

> I am not saying that climate change and SD subjects are not important, I am saying that there are more important issues for me as a faculty at the college to prioritise. I need to address the curriculum requirements and make sure I address my annual objectives with my students. I do not mind integrating these two important issues [climate change and SD concepts] into my teaching but I need a clear plan and learning outcomes prepared by the college. I will be able to do this after that (Participant 3)

The participants clearly stated that their lack of understanding of what is exactly expected from them in this respect, along with the lack of knowledge on how to integrate climate change and SD matters into their teaching, reveals a gap in understanding the support level that the faculty might need to proceed on such issues.

6.4.2 Climate Change and SD Matters Are not Seen as Important

The data showed that the participants believed that climate change and SD matters are not important for their courses. They expressed their concerns about linking such matters to their courses, which they thought were completely different disciplines. It appears that the faculty at the College has a gap in understanding that climate change and sustainable development are relevant to all their courses and that they can be used as part of teaching materials across various disciplines. For example, one faculty member stated: 'My course addresses different concepts … none of them are related to climate change. I just cannot see how to do this? Or I might be missing on something?' (Participant 2). There might be a role for the College to play here in order to guide the faculty on how to process such integration. This is mentioned by one of the participants as being:

> it is extremely important for such topics to be managed by the college in order for us [the faculty] to be able to implement them robustly, we need more motivation to deliver on such important matters, and such motivation can be implemented by the College (Participant 4)

The College needs to encourage the faculty to adopt such an attitude to promote it amongst their students. 'Institutions can recognise and address this opportunity by tapping into the strong movement to change, with student interest groups and faculty already taking steps to integrate these competencies' (Hackett et al. 2020: e2).

6.4.3 Students Are not Playing Part in Promoting Climate Change and SD Issues

The students are meant to be the catalysts of change in relation to sustainable development and climate change; however, the data showed that there is little role for the students at the College in addressing such challenges. It appears that the College and the faculty are not fully aware of what the students' role might be in this respect.

> I sometimes speak to my students about climate change and they are not come across as fully interested in this …. I do not, of course, speak to them in detail but sometimes we talk about news and activities around climate change … I can assure that my students do not seem to be fully comprehending this matter (Participant 2)
>
> I believe it is important for students to play a role in promoting sustainability and climate change issues, especially when they leave [the College] and go to schools. They need to be shown how to do this … also we [the faculty] need to be shown too (Participant 3)

The participants revealed that their students have not been active in promoting climate change and SD issues because there is currently no plan that can guide them on how to do so. This applies to the faculty who also seem incompetent in this field and who believe that there needs to be a clear policy that governs the practices of involving students and faculty in the process. 'If students are educated to understand and act upon climate changes, they can create a grassroots movement that produces systemic changes' (Remiers 2021: 153).

6.5 Conclusion and Recommendations

This paper argues that the College of Education at Qatar University can play an important part in promoting climate change and sustainable development matters amongst its faculty and students. The integration of climate change and SD matters into the teaching practice and curriculum can accomplish many benefits to help the State of Qatar achieve its 2030 vision in this respect. There are many benefits that such integration can achieve such as preparing the graduates of the College to work in a climate change environment, i.e. our world is facing many challenges in this respect and equipping the College's graduates with the necessary knowledge would make a difference by enabling them to be the catalysts of change. In addition, the faculty at the College can also promote eco-friendly teaching practices and integrate climate change topics into their subject matter by choosing climate change materials to teach the concepts of their disciplines. Finally, equipping the faculty and the students at the College with the necessary skills through training and updating of policies can strengthen the existing teaching practices, elevate the graduates' capabilities?, and enable the College to meet its responsibilities about climate change and sustainable development being the main educators' trainer in the country.

In light of the findings of this study, there is a need for the College of Education to be the lead in the country to raise awareness about climate change and sustainable development topics amongst its staff and students. This can be achieved by reviewing the curriculum designs and implementing new mechanisms in the delivery of teaching to meet the responsibilities that the College has towards the communities and the country. This paper recommends that the College focuses on reviewing and updating four main provisions to meet its responsibilities towards the environment.

6.5.1 Updating Learning Outcomes

Updating the learning outcomes to include equipping students with the necessary knowledge about climate change and sustainable development would enable them to be the catalysts of change in their workplaces. This needs to be addressed comprehensively by updating the faculty's knowledge and practices and raising students' awareness about the importance of such change for the future generation. In doing so, the College will be able to train educators who are accountable towards their society and capable of influencing the views and practices of the children they would teach about climate change and other important subjects.

6.5.2 Improving Teaching Practices

Exploring creative and innovative teaching methods that promote citizenship locally and globally would raise the future generations' awareness about the level of responsibilities that rests upon them to face the climate change challenges. Including climate change topics and adapting the teaching methods to highlight the climate change aspects can make a difference in raising both teachers' and students' awareness about the importance of this matter to the planet and to humanity in general. Practices designed to reflect climate change can include internship projects for students, dedicated lectures, research projects, tutorial sessions, debates and extra-curricular activities. Highlighting the importance of sustainability in running our schools and the behaviour change to meet our social obligations can prepare a generation that has a sound conceptual knowledge about appreciating what we have and working on protecting it.

6.5.3 Creating Effective Assessment

Creating effective alignments between the learning outcomes criteria and the assessment methods through a robust feedback system and a strong engagement of stakeholders can establish a comprehensive system of assessment that takes into consideration the development of the students in their study disciplines and the development of their knowledge related to climate change aspects. A well-rounded assessment system can define the College's priorities and enable students to prioritise their learning objectives. In order to achieve this, we need to make sure that our assessment criteria are related to our learning outcomes and are specific and realistic.

6.5.4 Reviewing Institutional Culture

To review the prevailing culture amongst students, faculty and staff if an effective climate change initiative is to be achieved and become successful. The College has a high potential to lead the change in the higher education sector in Qatar because of its role in providing schools with trained teachers, who can play a decisive role in raising the awareness of the children, who are our future generations. Introducing an interrelated culture between research and practice can change the perceptions of the College's population. Enabling and training the faculty to research climate change topics related to their disciplines and implementing their findings in their teaching practices would create a learning environment where students can flourish and understand the necessity of knowing about climate change.

References

Du Pisani JA (2006) Sustainable development–historical roots of the concept. Environ Sci 3(2):83–96

Galletta A (2013) Mastering the semi-structured interview and beyond: from research design to analysis and publication. NYU Press, New York

Hackett F, Got T, Kitching GT, MacQueen K, Cohen A (2020) Training Canadian doctors for the health challenges of climate change. Lancet Planet Health 4(1):e2–e3

Hajian M, Kashani SJ (2021) Evolution of the concept of sustainability. From Brundtland Report to sustainable development goals. In: Sustainable resource management. Elsevier, pp 1–24

Holden E, Linnerud K, Banister D (2014) Sustainable development: our common future revisited. Glob Environ Chang 26:130–139

Parra G, Hansmann R, Hadjichambis AC, Goldman D, Paraskeva-Hadjichambi D, Sund P, Conti D (2020) Education for environmental citizenship and education for sustainability. In: Conceptualizing environmental citizenship for 21st century education, pp 149–160

Redclift M (2005) Sustainable Development (1987–2005): an oxymoron comes of age. Sustain Dev 13:212–227

Reimers FM (2021) Education and climate change: the role of universities. Springer Nature, p 201

Samalisto K, Lindhvist T (2008) Integration of sustainability in higher education: a study with international perspectives. Innov High Educ 32:221–233

Vaismoradi M, Jones J, Turunen H, Snelgrove S (2016) Theme development in qualitative content analysis and thematic analysis. J Nurs Educ Pract 6(5):100–110

Van Zon H (2002) Geschiedenis & duurzame ontwikkeling. Duurzame ontwikkeling in historisch perspectief: enkeleverkenningen. Werkgroep Disciplinaire Verdieping Duurzame Ontwikkeling, Nijmegen/ Groningen

Veblen T (1917) The technicians and the revolution. Reprinted in: The portable Veblen. Viking Press, New York

Mayamin Altae is an Assistant Professor at the Faculty of Education, University of Buckingham (UK), with over 20 years of experience in teacher training, research, and higher education across the UK and the MENA region. She holds a PhD in Education from the University of Leicester and an MA in Educational Technology and TESOL from the University of Manchester, and is deeply committed to integrating technology into education. As Head of Initial Teacher Training International Courses, she also advises governments and organisations across the Middle East on advancing the UN Sustainable Development Goals—particularly Quality Education (Goal 4) and Climate Action (Goal 13)—and has published several papers addressing key educational challenges.

Connecting Sustainability and Global Citizenship in Universities in the Gulf Cooperation Council

7

Wendy A. Booth

Abstract

Sustainability is often framed in relation to environmental issues alone, yet the global indicator framework for the Sustainable Development Goals (SDGs) calls for more education in three key areas: global citizenship, sustainable development and climate change. However, it has been found that policy makers and educators, including in the Gulf Cooperation Council (GCC) countries, sometimes struggle to address the global citizenship element. Therefore, finding ways to support educationalists and higher education institutions in promoting global citizenship and social justice is essential. This paper presents a scoping review that examines the efforts made by the countries of the GCC to advocate education for sustainable development and global citizenship, particularly within higher education, and puts forward recommendations for teaching and developing resources that are culturally and contextually appropriate.

Keywords

Global citizenship · Higher education · Sustainability

W. A. Booth (✉)
University of South Wales, Cardiff, UK
e-mail: wendy.booth2@southwales.ac.uk;
wendy@t-teema.com

7.1 Introduction

The global indicator framework for the Sustainable Development Goals (*SDGs*) includes a call for more education in three key areas: global citizenship, sustainable development and climate change. Notably, UNESCO integrates both education for Sustainable Development (ESD) and Global Citizenship Education (GCE) under its education agenda. ESD equips learners with knowledge and skills to promote sustainability (e.g. climate action, resource conservation), and GCE fosters awareness, critical thinking and active participation in addressing global issues such as inequality, peace and human rights. Together, they aim to empower individuals to become informed, responsible and active global citizens who contribute to the 17 Sustainable Development Goals (SDGs). Education for sustainable development and global citizenship (ESDGC), then, prepares individuals to think critically, act responsibly and contribute meaningfully to a better world.

While sustainable development and global citizenship encompass many of the problems being faced by societies around the world today, an examination of policies and approaches used for teaching these concepts is required to ensure impact. Students around the world must properly understand these concepts of sustainable development and global citizenship to support them in becoming better citizens able to advocate for

K. Al'Abri and E. Buckner (eds.), *Higher Education and Sustainable Development in the Arab Gulf States*, Sustainable Development Goals Series, https://doi.org/10.1007/978-3-032-19559-3_7

positive change that benefits others; moreover, consideration of different cultural and contextual factors is needed to both analyse and develop educational approaches. Therefore, this paper is an attempt to bridge the gap in the knowledge with regard to teaching global citizenship in the specific cultural context of the Gulf Cooperation Countries; it explores both the strengths and shortcomings of policies on education related to the SDGs, leading to recommendations for teaching and developing effective resources that are culturally appropriate.

According to Aleixo et al. (2016), sustainable development is made up of three key dimensions, namely, environmental, economic and social. Relatedly, the United Nations' 17 Sustainable Development Goals (SDGs) address a range of areas related to global citizenship, and not just matters related to the environment and climate change; for example, zero hunger (Goal 2), gender equality (Goal 5) and quality education (Goal 4), which highlights a clear link between the SDGs and global citizenship. This link can be seen in education policy in countries such as the UK, where Education for Sustainable Development (ESD) comes with the addition of Global Citizenship (ESDGC); GC is typically conceptualised as a liberal approach to citizenship that aims to make learners responsible citizens, encompassing civil and social rights (Jenkins 2021). However, it has been found that while educators feel comfortable discussing climate change and environmental issues, they sometimes struggle to address sustainability in relation to the citizenship aspect of the SDGs, perhaps because it can lead to some difficult and controversial questions. Therefore, this paper will examine global citizenship in the context of the Gulf countries and explore ways of supporting educators of ESDGC to encourage critical thinking, empathy and tolerance as part of a sustainability agenda.

The importance of ESDGC is reiterated by Akkari and Maleq (2020), and they explain that "efforts to move along a sustainable development path may only be achieved by promoting global social justice" (p. 3). Related to this, an area of global citizenship that can be difficult to teach is tolerance and ethnocultural empathy, despite these being key to fair notions of citizenship and social justice, especially as part of anti-racist and anti-bullying strategies. Notably, these issues fall under Principle Two of the agenda of the SDGs: leave no one behind—an agenda of the 17 goals that includes equality and ending discrimination (UNSDG 2024). Related to this, Dreamson (2018, p. 78) explains, 'Being a global citizen means that the learner is capable of emphasising and putting oneself in other people's shoes; developing responsibility in local, national and global context; and contributing to the wider world community'.

While governments and organisations such as the UN use the term global citizenship, Massaro (2022) explains that it has received criticism because the word 'citizen' refers to being a member of a state, and on the global level, individuals do not have the rights offered to them in the same way as being part of a state. Even so, global citizenship, like other terminology that attempts to envisage difficult concepts, can be adapted, leading to perhaps more accurate terms such as 'global mindedness, globally-minded, global learner and globally oriented. These terms pull away from the idea of citizenship and put forth the notions of knowledge, understanding, consideration and empathy' (Massaro 2022, p. 99). Along with defining what exactly global citizenship means, putting it into practice within educational settings presents a myriad of problems and potential solutions. With this in mind, this paper explores perspectives on global citizenship, government policy and institutional aims and potential teaching methods and resources suitable for use within the GCC.

In the context of the Gulf Cooperation countries, Zguir et al. (2021) suggests learning from and collaborating with other countries and models in an attempt to meet all of the United Nations' Sustainable Development Goals (SDGs). One approach used within social science disciplines typically within a Western context is to teach students about ethical philosophy. However, when considering this in the context of the Gulf countries, consideration of culture and religion is required; hence, this paper examines some culturally and religiously suitable potential options, including drawing on social psychology and sociological theory as a way of encouraging critical thinking.

Social psychology can be incorporated into global citizenship education (essentially social justice education) to provide an understanding of why people behave as they do, which can support greater ethnocultural empathy and tolerance towards others. For example, the tendency of people to view their own 'in-group' in a positive light while viewing the 'out-group' negatively, as posited by scholars such as Levine et al. (2005), provides a useful way of examining conceptions of 'the other'. Added to that, sociological theories such as structural inequalities can be used to explore social problems and potential solutions. Such seemingly abstract concepts, combined with real-life examples, can promote positive citizenship and support critical thinking skills. These topics lend themselves to the social sciences, but within an inclusive higher education environment, staff training and extra-curricular courses can provide a way of extending global citizenship to other departments and subject areas.

In summary, global citizenship and social justice are key to achieving sustainable development, and along with policy initiatives, providing educators with training in this area is essential to supporting their efforts and working towards achieving the SDGs. Glover et al. (2013) explain that higher education students are our future leaders; therefore, it is essential that they appreciate the social implications of their current and future work, and the challenges around sustainable development and global citizenship. They state: 'students must, as part of their higher education experience, be equipped with the social and ethical context of their discipline and develop the critical thinking, analytical and communication skills that will ensure they can understand and approach with sophistication the environmental and social context within which they work' (Glover et al. 2013, p. 76). To ensure that the SDGs are met, and as each country commits to the SDGs in its own unique way, clear government guidelines and policies need to be put in place to guide educators. Therefore, the next sections form a scoping review of government policy and guidelines, and what is happening in the Gulf Countries, specifically in relation to education for sustainable development and global citizenship.

7.2 Literature Review

Governments around the world have implemented various policies to meet the SDGs, including within the field of education; however, while policy makers may have a clear idea of what needs to be done, educators' understandings of this may vary. As Braun et al. (2010) explain, policy enactment is the 'understanding that policies are interpreted and 'translated' by diverse policy actors in the school environment, rather than simply implemented' (p. 547). Moreover, as policy is implemented within a variety of contexts, the way it feeds down into practice will vary. That is, while governments legislate policy, its enactment involves 'negotiation, contestation or struggle between different groups who may lie outside the formal machinery of official policy-making' (Ozga 2000, p. 113 cited in Maguire et al. 2015, p. 2). Regarding government policy and the enacting of a curriculum, Sherrington (2017) describes four dimensions, the first two of which are the intended curriculum and the enacted curriculum, suggesting that what is intended by governments is not always enacted. Added to that, education policy shifts over time, as Maguire et al. (2015, p. 3) explain:

> Policy is not 'done' at one point in time, it is always a process of 'becoming'. It is reviewed and revised as well as sometimes dispensed with or sometimes simply just forgotten. There will be multiple subjectivities and positions that will shape how policies are understood, and differences will occur in enactments over time and in different spatial contexts. Enactment then is messy, incomplete and a form of interpretation and intersubjectivity in action.

Similarly, and with regard to the Gulf Countries in particular, Hayes and Findlow (2020 p.191) explain that 'Because of time-sensitive policy priorities, the original global–local relationships in a nation state, established within domains of geographies, shared histories, cultures and political ideologies, do not remain constant'. Furthermore, Abubakar et al. (2020) describes the increasing attention that sustainable development has been receiving within higher education over the past three decades, yet while an institutional framework is fundamental to advancing sustainable development, 'some universities, especially in the countries of the Global South, are either missing this dimension or have inefficient institutional frameworks due to the lack of standard practice or guidance' (p.1). Therefore, the next section explores some of the government policies on sustainable development, particularly global citizenship, in the countries of the Gulf Cooperation Council in an attempt to unpack what is happening and what needs to be done to move this agenda forward.

The Gulf Cooperation Council is a regional intergovernmental and economic union made up of (in alphabetical order): Bahrain, Kuwait, Oman, Qatar, Saudi Arabia and the United Arab Emirates. This section provides an overview of some of the unique issues faced, and efforts being made, by the various GCC countries to integrate the SDGs into their respective curricula. Research by Ullah et al. (2020) shows that anti-migrant sentiments are on the rise in the GCC, making education for global citizenship and social justice crucial. In addition, Zaidan et al. (2019) describe how the Gulf Countries are engaging an increasingly global audience and receiving more visitors, with mega events such as the 2020 World Cup held in Qatar adding to the push towards tourism and knowledge-based industries to diversify the economies of the region (Hayes and Findlow 2020; Abubakar et al. 2020; Ahmed 2018). This is leading to international scrutiny of environmental standards and social justice related to the sustainability agenda. Therefore, Zaidan et al. (2019) state that education efforts are needed in schools, universities and local communities to create more awareness of sustainable practices and for individuals to realise their own responsibilities regarding sustainable development, including:

> integrating messages within school curricula, investing in electives in universities across disciplines, examining the Islamic perspective on sustainability and encouraging further research and education around this, and engaging community leaders and influential personalities to advocate for sustainable development (p.677).

In this way, effective policy can be developed and transformed into practice, helping to achieve the SDGs and the global indicator framework of more education in the areas of global citizenship, sustainable development and climate change.

In Saudi Arabia, major transformations are taking place based on the country's Vision 2030 agenda, with the government's Vision 2030 (2023) website stating, 'Sustainability has been at the heart of Vision 2030 since its inception'. However, much of the focus is on green issues and a number of outstanding environmental projects. On the other hand, a more encompassing approach to sustainable development has been taken by the Ministry of Education in Saudi Arabia, as the ministry website states it is working in multiple areas towards: 'Achieving societal sustainability; Achieving economic sustainability; Achieving environmental sustainability'. In addition, the Ministry of Education's Sustainable Development Goals include the following points: 'Encouraging continuous education and affording opportunities to all' and 'Continuing to act upon the principle of gender-fair and balanced access and opportunity in education' (Ministry of Education 2023).

Abubakar et al. (2020, p. 2) claim that Saudi Arabia has embarked on 'incorporating SD into their university systems especially in campus operations and programmes to better equip students to deal with SD challenges in all life endeavours and contribute in practice with far-reaching impacts on both local and global futures'. However, Abubakar et al. (2020) highlight environmental causes, rather than social issues, such as water efficiency and waste reduction; therefore, they recommend training future decision makers on the importance of sustainability, including 'social equity'. Similar to other countries, Al-Dosary (2016) found that the education system in Saudi Arabia is better equipped at education for sustainable development from a natural sciences perspective; therefore, they claim that the curriculum has failed to live up to the Saudi government's commitment to sustainable development and global citizenship. Furthermore, Algraini (2021, p. 421) found that 'Saudi education tends to focus on economic benefits and gives little consideration to the intrinsic value of education for whole-person development and lifelong learning'. Based on this, they suggest encouraging critical thinking and problem-solving skills to allow students to engage in innovating and analysing modern-day issues (Algraini 2021).

The United Arab Emirates has focused on a key element of global citizenship through its promotion of the notion of tolerance, with 2019 declared the Year of Tolerance in the country (Al Majali and AlKhaaldi 2020). Although the term tolerance is contested by some due to the idea that it means putting up with something one dislikes, its usefulness lies in its simplicity, and its common usage (Booth 2025); for example, in the UK, the Department for Education (2014 p. 5) in England states: 'Schools should promote the fundamental British values of democracy, the rule of law, individual liberty and mutual respect and tolerance of those with different faiths and beliefs', or as Verkuyten and Yogeeswaran (2017, p. 73) put it: 'A diverse, egalitarian and peaceful society does not require that we all like each other, but it does require that people at least tolerate one another'.

Furthermore, the We the UAE 2031 Vision clearly considers societal factors, as it states that its focus is on 'social, economic, investment and development aspects'. This is based on the following four pillars: firstly, 'Forward Society—achieving the prosperity of society by enhancing the capabilities of the citizens to maximise their effective contribution in all sectors'. Secondly, 'Forward Economy—reflecting the UAE's belief in the importance of human capital as the main driver of the next 10-year development plan'. Thirdly, 'Forward Diplomacy—consolidating the pivotal role and influence of the UAE based on respect for human values' and fourthly, 'Forward Ecosystem—enhancing the government performance and the UAE's infrastructure and its development according to the latest technological methods, including the development of digital infrastructure (The United Arab Emirates' Government portal 2023).

Al-Naqbi and Alshannag (2018) investigated United Arab Emirates University students' knowledge, attitudes and behaviours towards education for sustainable development, using a cross-sectional survey. Eight hundred and twenty-three students responded, and it was found that they had high levels of understanding of, and positive attitudes towards, education for sustainable development. However, while the results are positive, the research focused on environmental, rather than societal, factors, which it seems is typically the focus in the region when sustainable development is on the agenda. More specifically, Wren (2021) researched global citizenship education among 50 Emirati students using a quasi-experimental design that included five weeks of teaching citizenship as part of sessions on social responsibility, which had previously focused mainly on volunteer work. They found that 'the global citizenship curriculum advanced students' understanding of social responsibility and civic concepts, specifically civic awareness, social justice and diversity'. Therefore, they suggest that 'educators should experiment with curriculum redesign to further facilitate the development of social responsibility in undergraduate students' (Wren 2021, p. 80). Importantly, Wren (2021) implemented a curriculum based around the students' global orientation, empathy and cultural sensitivity, rather than using a Western-based approach, and this facilitated consideration of cultural and religious values. Hence, the study shows that students are accepting of the topic of global citizenship and that designing teaching resources relevant to the local environment can support 'respect for human values' (The United Arab Emirates' Government portal 2023).

Qatar's National Vision 2030 (Government Communications Office 2023) highlights 'the importance of knowledge acquisition, production and dissemination and promotion of human socio-economic development' (Ahmed 2018, p. 39). Furthermore, Qatar's Vision 2030 is based on four main pillars, which are:

1. Development of all Qatar's people to enable them to sustain a prosperous society.
2. Development of a just and caring society based on high moral standards and capable of playing

a significant role in the global partnership and development.
3. Development of a competitive and diversified economy capable of meeting the needs of and securing a high standard of living for all its people for the present and future.
4. Management of the environment such that there is harmony between economic growth, social development and environmental protection (Government Communication Office State of Qatar 2023).

It can be seen that societal, as well as environmental and economic factors, feature in the vision. Again, the issue is: how can such pillars/policies/aims be translated into practice. An important issue to consider is that the 'preservation of cultural traditions is a major challenge that confronts Qatar knowledge society in a rapidly globalised and increasingly interconnected world' (Ahmed 2018 p.50). Similar to the other GCC countries, important traditional social values such as generosity and tolerance, 'have been challenged by the globalisation of greater freedoms, wider choices and advancing individualism and personal interests that accompany economic and social progress in the modern world' (Ahmed 2018 p. 50). This is an interesting point in that traditional values that form a key part of education for global citizenship are being negatively impacted by globalisation and so-called social progress, which raises questions around how global citizenship is implemented and whose values exactly should take precedence to secure a sustainable future.

The Kuwait Vision 2035 contains three key visions that end with 'consolidating the values of society, preserving its identity, as well as achieving justice, political participation and freedoms' (Ministry of Foreign Affairs 2023), which again emphasises the importance of the country's own norms and values. Furthermore, Oman Vision 2040 (2023) includes citizenship, identity, heritage and national culture, yet different to some other countries, it has 'set its strategic direction in conjunction with a society proud of its identity and culture and committed to its citizenship' (Al'Abri et al. 2022, p. 11). This highlights the national and local elements of citizenship, rather than the global. Along a similar vein, Hayes and Findlow (2020, p. 190) explain that Bahrain has taken a 'glocalised' approach to policy-making, incorporating both local needs and global influences, which has led to a culturally authentic 'unique HE strategy'. While a great deal of focus is on advancing the country regionally, policy documents include strategic decisions that capture a sense of the wider world. The focus on the local includes the launch of Bahrain Economic Vision 2030, which 'aims to create a more prosperous and secure future for Bahraini citizens and residents' (Bahrain BH 2020), which is based on the principles of 'sustainability, competitiveness and fairness in the economy'. Furthermore, Priority 1 of the Government Action Plan (2019–2022) includes 'strengthening the democratic system and human rights' (Bahrain SDGs 2030 2022).

AlMaamari (2020 p.72) explains that teaching global citizenship to university students may be seen as controversial due to the variety of perspectives on what specific attributes should be promoted. While Article 21 of the Basic Law in Oman states: 'Citizens are all equal before the law, and they are equal in rights and public duties, and there was no discrimination between them based on gender or origin, colour, language, religion, sect, domicile, or social status' (Royal Decree 2021 cited in Al'Abri et al. 2022, p. 11), situating this within the curriculum may be complex. To explore this, AlMaamari (2020) conducted research at the College of Education in Oman, which involved a 13-week course for 49 students on global citizenship that covered a range of topics, including peace, tolerance and acceptance of others; values of justice; human rights and global citizenship from an Islamic perspective. They found that the students who completed the course 'tended to identify themselves as global citizens by showing positive perceptions of GC, especially in terms of cognitive and socioemotional dimensions' (AlMaamari 2020 p.81). In addition, the participants acknowledged the importance of helping and donating to others as part of their religious duty and responsibility as global citizens. Therefore, AlMaamari (2020) claims that further research is needed to obtain the views of policy makers and curriculum developers in order to form more cohesive approaches to global citizenship.

The need for appropriate ways of educating for global citizenship is posited by Bauer et al. (2018, p. 497) as they state that 'Processes and structures for sustainability governance cannot simply be copied without re-contextualising them for adaptation by other university campuses' and this point is even more complex when it comes to the cross-cultural adaptation of global citizenship curricula. As shown above, although there are some shortcomings, the governments of the GCC have introduced policies that relate to sustainability and global citizenship; however, translating these into practice can be difficult. Therefore, the next section presents some practical solutions for teaching global citizenship and social justice in the context of the GCC, to ensure that local cultural and religious norms are respected, while addressing key points related to the SDGs and the global indicator framework.

7.3 Discussion and Analysis

7.3.1 Practical Solutions to Teaching Global Citizenship Cross-Culturally

The previous section discussed education for sustainable development and global citizenship in the context of the GCC countries. While government policy focuses on many of the key issues related to EDSGC, it is important to ensure that this policy filters down into education settings. Furthermore, culturally appropriate models are needed to ensure the effectiveness of global citizenship education and to draw on the positive attributes that already exist within religious and cultural contexts. This recognition of a global citizenship curricula that goes beyond Western conceptions is discussed by Dreamson (2018, p. 75), and he argues that 'culturally inclusive GCE needs to be reconceived in a way to promote authentic intercultural interaction, and to do so, non-Western cultures/religions need to be repositioned as equal participants in culturally inclusive GCE development'. Therefore, this section provides an overview of key concepts and suggests ways in which social psychology, sociology and critical thinking skills can be taught as part of a global citizenship agenda that encompasses a wide range of societal issues while remaining culturally sensitive.

7.3.2 Social Psychology for Teaching Empathy

Baron-Cohen (2011, p. 11) defines empathy as follows: 'Empathy begins when we suspend our single-minded focus of attention and instead adopt a double-minded focus of attention'; put simply, it means walking in the shoes of others. Empathy is a useful concept from a cross-cultural perspective as it can be taught and encouraged in a way that does not focus on a particular ethnicity, 'race', religion, sect or culture, as advocated by Booth (2017); rather, it focuses on group dynamics; in particular, attitudes towards in-groups and out-groups and the idea of 'the other' in society (Levine et al. 2005).

There are four main approaches to empathy according to Batson and Ahmad (2009), which are the imagine-self perspective; the imagine-other perspective; emotion matching and empathic concern. The imagine-self perspective refers to the person imagining how they would think and feel if they were in someone else's situation, and it can be traced as far back as Adam. U Smith (1759/1853) who described it as 'changing places in fancy'. Research by Wren (2021) involving first-year female university students in the UAE found that a global citizenship education curriculum led to improvements in the social responsibility subscales of civic awareness, social justice attitudes and diversity attitudes, as well as empathy. Taking an imagine-self perspective can reduce stereotyping of an out-group by imagining being in the position of someone from that stigmatised out-group, and it requires the engagement of cognitive processes, such as attempting to understand behaviour.

An imagine-other perspective is sensitive to the feelings and desires of the other person, and an example of its impact is shown in the research by Vescio et al. (2008). The research sample was White undergraduate students at a US university, and they were required to watch a video of a young Black man describing the racial discrimination he has faced. Following the video, the students were asked to complete a questionnaire using Katz and Hass's (1988) pro-Black and anti-Black attitudes scale, which assessed their emotional responses to the video; the attributions of man in the video and the extent of his responsibility for what happened and their general attitudes towards Black people. Vescio et al. (2008) found that, compared to a control group, the students who attempted to take an imagine-other perspective attributed the racism suffered by the man more to situational causes than the control group, as well as having more positive attitudes towards Black people and greater empathic concern for the young Black man in the video.

But why is this important in the context of the GCC? Empathy and understanding others are essential when it comes to treating others well, including migrant populations that have moved to the region (Ullah et al. 2020). It is important to encourage philanthropic attitudes in an era when climate change is having devastating consequences around the globe and adding to increasing rates of migration. On a local level, ethnocultural empathy is important in understanding others and the smooth functioning of society; for example, in the field of healthcare, Rasoal et al. (2009, p. 15) explain that it is 'at least plausible that a person with little or no empathy for persons from a different cultural background will have problems in patient consultations'. Furthermore, the United Nations (2016) has been using virtual reality to inspire humanitarian empathy and encourage people to support government efforts to implement the SDGs and alleviate poverty. This use of virtual reality is linked to the idea of attempting to walk in the shoes of others, as it allows users to virtually experience the suffering of others and consider both their life circumstances and life choices.

Therefore, it can be concluded that it is possible to teach how to be more empathic by encouraging alternative perspectives and using video footage to learn about the experiences of others.

7.3.3 Sociological Concepts

C. Wright Mills' concept of the sociological imagination facilitates 'the capacity to shift from one perspective to another—from the political to the psychological' (Mills 1959, p. 7). It involves imagining walking in the shoes of others, but combined with an awareness of historical and social factors, and consideration of how attitudes and feelings towards others are formed. Phillips (2015, p. 48) explains how this requires 'examining the ways in which individuals are affected by broader meso-level sociological phenomena'. Thus, it involves empathy as part of social justice, while highlighting how structural factors impact the lives of others and affect their life chances, their mindset and their opportunities for growth. This can support learners not only to change their attitudes to be more positive towards others (e.g. the disadvantaged and refugees), but also to want to come up with solutions—a key agenda of the SDGs. However, there are limitations to this approach, as attempting to look at life from another perspective is not an easy feat, and while people may be aware of some issues in their own contemporary society, at the same time they may be unaware of important economic and social factors at play in other contexts. Even so, trying to take the perspective of others is important and not something that should be ruled out because of these shortcomings, and research conducted by Booth found that promoting their sociological imagination can have a positive impact on learners' attitudes. From a practical pedagogical perspective, this topic provides an opportunity for class discussion and critical debates that should lead to greater awareness of the structural factors that exist in society and opportunities that improve or stifle individuals' life chances. In this way, a sense of global citizenship can be fostered.

7.3.4 Practical Pedagogy

The previous sections have explained how social psychology and sociological concepts can be useful to fostering global citizenship, or global mindedness (Massaro 2022, p. 99), and so this section attempts to put forward suggestions for the more practical teaching of these concepts to support educators who may not be entirely familiar with them. Important to note is that in the context of good citizenship, moral educators are required to 'cultivate in children and young people the sort of habits, desires, dispositions and inclinations that will move them to do what morality requires' (Hand 2017, p. 2). However, research has shown that teachers sometimes struggle with concepts related to the global citizenship aspect of sustainable development (Al'Abri et al. 2022; Al-Thani et al. 2021; AlMaamari 2020; Glover et al. 2013).

A starting point for encouraging learners to think about global citizenship; social justice and inequalities in society, including those related to the SDGs such as zero hunger, gender equality and quality education (United Nations 2023), is to ask about their own position in society—the family they were born into, their gender, whether they have a disability, their wealth and social class and so on. In line with Rawls (1971) question how they managed to be born into that situation, and they will realise that they had no choice in the matter. Hopefully, this will be a humbling experience, and in the context of Muslim societies, this is in line with the widely shared advice of Prophet Muhammad (SAWS): 'Look at the person who is below you, and do not look at one who is superior to you. Thereby you would be able to better appreciate the favours that Allah has bestowed upon you'. Next, ask learners to imagine their lives without the advantages or disadvantages they have experienced, for example, would they have been able to complete their studies? It is useful to begin this task on an individual basis and then extend it to a class discussion by dividing learners into small groups. In this way, they can draw on their sociological imagination to share ideas and reflect as a form of critical thinking.

Global citizenship is an interdisciplinary subject that is not only relevant to the social and political sciences. It can be embedded into history, as if we know the historical experiences of groups in society, we can better understand their current situation, such as the experiences of refugees and third-generation migrants who continue to be impacted by what went on before them. In geography, the push–pull factors of migration, the impact of climate change and the idea of a world with and without borders provide a wide range of points for discussion. From a sociological perspective, structural factors (social, economic and institutional), as well as inequalities and notions of (un)fairness, provide important starting points for critical thinking. Other useful theories that can be used are labelling theory and self-fulfilling prophecy (Merton 1948), which refer to how the self-identity and the behaviour of individuals may be influenced by the words used to describe or classify them (Merton 1948); in the long term, this can affect the way individuals perceive themselves, with the choices they make in life based on this self-perception. Students can be asked if they have ever been labelled themselves as a starter activity.

Osler (2015) advocates human rights as a basis for education and teaching pupils to challenge injustice, racism and xenophobia; however, teachers may not have the background or training necessary to address such topics effectively and they may not feel comfortable doing so. To mitigate this, it can be seen that in the pedagogical approaches above, no specific group in society has been used as an example, which allows the teacher to guide the focus towards issues they are comfortable discussing. Despite this, introducing real-life examples through the use of film or virtual reality can be useful in fostering empathy, such as exploring fast fashion and consumerism, which is itself an important aspect of sustainable development, by watching documentaries on 'sweatshop' factories and the lives of people working there. Essential to critical thinking is giving students points to take notes on while watching video content, rather than allowing them to do so passively. Engaging students through activities and discussion tasks should result in a more engaging and impactful experience.

7.4 Conclusion and Recommendations

According to Ahmed, 'the education system has to encourage analytical and critical thinking, as well as creativity and innovation. It has to promote social cohesion and respect for society's values and heritage and advocate for constructive interaction with other nations' (Ahmed 2018, p.48). In addition, Dreamson (2018, p. 90) suggests 'seeking cultural alliances of non-Western cultures/religions for developing culturally inclusive GCE pedagogical strategies and values while continuing to be critical to epistemic violence/othering processes'. With regard to teaching empathy in the context of the Muslim-majority countries of the Gulf region, the infamous story of the man who was forgiven by God because he put water from a well in his shoe to enable a thirsty dog to drink, provides an example of how empathy is embedded into belief and a suitable concept within GCE.

The gaps that continue to exist between policy and practice in education for sustainable development (Amin et al. 2023) are not unique to the Gulf Cooperation countries; for example, a European Commission (2021) report highlights a gap between policy strategies and how these are put into practice across countries in Europe. To address the disconnect between policy and what actually happens in education institutions, and based on their research in Saudi Arabian universities, Abubakar et al. (2020) suggest introducing an institutional framework made up of formal policies and procedures and informal practices, starting with a vision for sustainability. In addition, social science subjects should receive more attention both through improving teaching practices in that field and in providing support to other subject areas so that all students become aware of the importance of 'shared responsibility for solving global issues' (Carabain et al. 2012, p. 30). The research conducted by Wren (2021) in the UAE shows that engaging students in global citizenship education through reflections and diverse dialogue taught them to respect differing viewpoints; explore a range of perspectives, cultures and worldviews and consider how they can serve their community.

The process of continual improvement in GCE is ongoing, especially as 'The GCC context has undergone sweeping educational reforms and is still attempting to find a balance between localisation, preservation of tradition and language as well as being competitive in a globalised world' (Amin et al. p.1). Education for sustainable development and global citizenship requires a wide range of skills that encompass environmental, economic and sociological aspects. Therefore, based on this chapter's exploration of global citizenship education in the Gulf Cooperation countries, I make the following recommendations to educators and universities:

- Ensure that government policies are filtering down into practice by ensuring that universities have a framework in place that supports education for sustainable development and global citizenship, including an appropriate action plan for the curriculum.
- Research and trial different approaches to teaching global citizenship and social justice, such as the use of social psychology and sociological constructs and theories to promote tolerance, ethnocultural empathy and a willingness to engage in positive change.
- Provide teacher training on global citizenship and social justice so that educators are equipped with the confidence and skills necessary to engage students and promote their critical thinking skills.
- Seek support from educators and researchers in the social sciences to disseminate relevant knowledge to other disciplines, decision makers and policy developers.

References

Abubakar IR, Aina YA, Alshuwaikhat HM (2020) Sustainable development at Saudi Arabian universities: an overview of institutional frameworks. Sustainability 12(19):8008

Ahmed FBJ (2018) Challenges of the knowledge society: exploring the case of Qatar. Glob Econ Obs 6(1):39–54

Akkari A, Maleq K (2020) Global citizenship education: recognizing diversity in a global world. Global citizenship education: critical and international perspectives. Springer International Publishing, Cham, pp 3–13

Al Majali S, AlKhaaldi K (2020) Values of tolerance in relation to academic achievements, cultures, and gender among UAE universities students

Al'Abri KM, Ambusaidi AK, Alhadi BR (2022) Promoting Global Citizenship Education (GCED) in the sultanate of Oman: an analysis of national policies. Sustainability 14(12):7140

Aleixo AM, Azeiteiro UM, Leal S (2016) Toward sustainability through higher education: sustainable development incorporation into Portuguese higher education institutions. In: Challenges in higher education for sustainability, pp 159–187

Algraini S (2021) Education for human development: a capability perspective in Saudi public education. Compare: J Comp Int Educ 51(3):416-432.

AlMaamari SN (2020) Educating for 'glocal' perspectives at Sultan Qaboos University: studying students' perceptions after studying a global citizenship education course. Citizen Soc Econ Educ 19(1):69–84

Al-Naqbi AK, Alshannag Q (2018) The status of education for sustainable development and sustainability knowledge, attitudes, and behaviors of UAE University students. Int J Sustain High Educ

Al-Thani WA, Ari I, Koç M (2021) Education as a critical factor of sustainability: case study in Qatar from the teachers' development perspective. Sustainability 13(20):11525

Amin H, Zaman A, Tok E (2023) Education for sustainable development and global citizenship education in the GCC: a systematic literature review. Global Soc Educ 1–16

Bahrain SDGs 2030 (2023) Bahrain's efforts to achieve sustainable development goals 2030. Available at https://www.sdgs.gov.bh/

Bahrain BH (2022) Sustainable development. Available at https://www.bahrain.bh/new/en/sdgs_en.html

Baron-Cohen S (2011) Zero degrees of empathy. Penguin, London

Batson CD, Ahmad NY (2009) Using empathy to improve intergroup attitudes and relations. Soc Issues Policy Rev 3(1):141–177

Bauer M, Bormann I, Kummer B, Niedlich S, Rieckmann M (2018) Sustainability governance at universities: Using a governance equalizer as a research heuristic. High Educ Policy 31:491–511

Booth W (2017) An interdisciplinary approach to the high school curricula: incorporating elements of social psychology into the Citizenship curriculum to increase levels of tolerance and Ethnocultural empathy. In Anastasiou N, Faustmann H, Hadjipavlou M, Karahasan H, Psaltis I, Zackheos M (eds) The role of education in a multicultural cyprus. Cambridge Scholars Publishing, Cambridge

Booth WA (2025) Educating for Community Cohesion: Promoting Tolerance and Ethnocultural Empathy. Palgrave Studies in Global Citizenship Education and Democracy. Palgrave Macmillan.

Braun A, Maguire M, Ball SJ (2010) Policy enactments in the UK secondary school: examining policy, practice and school positioning. J Educ Policy 25(4):547–560

Carabain C, Keulemans S, vanGent M, Spitz G (2012) Global citizenship public support for participation. NCDOPress, Amsterdam

Commission E (2021) Education for environmental sustainability: policies and approaches in European union member states. European Commission, Brussels

Department for Education (2014) National curriculum in England: framework for key stages 1 to 4 Available at https://www.gov.uk/government/publications/national-curriculum-in-england-framework-for-key-stages-1-to-4/the-national-curriculum-in-england-framework-for-key-stages-1-to-4. Accessed 20th August 2022

Dreamson N (2018) Culturally inclusive global citizenship education: metaphysical and non-western approaches. Multicult Educ Rev 10(2):75–93

Glover A, Jones Y, Claricoates J, Morgan J, Peters C (2013) Developing and piloting a baselining tool for education for sustainable development and global citizenship (ESDGC) in Welsh higher education. Innov High Educ 38:75–86

Government Communications Office (2023) Qatar National Vision 2030 https://www.gco.gov.qa/en/about-qatar/national-vision2030/

Hand M (2017) A theory of moral education. London: Routledge.

Hayes A, Findlow S (2020) The role of time in policy-making: a bahraini model of higher education competition. Crit Stud Educ 61(2):180-194.

Jenkins A (2021) Ecological global citizenship education: a reframing. Int J Develop Educ Global Learn 13(2):94–108

Katz I, Hass RG (1988) Racial ambivalence and American value conflict: Correlational and priming studies of dual cognitive structures. J Pers Soc Psychol 55(6):893.

Levine M, Prosser A, Evans D, Reicher S (2005) Identity and emergency intervention: how social group membership and inclusiveness of group boundaries shape helping behavior. Pers Soc Psychol Bull 31(4):443–453

Maguire M, Braun A, Ball S (2015) Where you stand depends on where you sit: the social construction of policy enactments in the (English) secondary school. Discour Stud Cult Polit Educ 36(4):485–499

Massaro VR (2022) Global citizenship development in higher education institutions: a systematic review of the literature. J Global Educ Res 6(1):98–114

Merton RK (1948) The self-fulfilling prophecy. The Antioch Rev 8(2):193-210.
Mills CW (1959) The sociological imagination. Penguin, Harmondsworth
Ministry of Education (2023) Sustainable development. Available at https://moe.gov.sa/en/aboutus/aboutministry/Pages/sustainabledevelopment.aspx
Ministry of Foreign Affairs (2023) Kuwait Vision 2035 "New Kuwait". Available at https://www.mofa.gov.kw/en/kuwait-state/kuwait-vision-2035/
Osler A (2015) Human rights education, postcolonial scholarship, and action for social justice. Theory Res Soc Educ 43(2):244–274
Ozga J (2000) Policy research in educational settings: Contested terrain. Open University Press, Buckingham
Phillips BS (2015) Understanding terrorism: building on the sociological imagination. Routledge, London
Rasoal C, Jungert T, Hau SE, Stiwne E, Andersson G (2009) Ethnocultural empathy among students in health care education. Eval Health Prof 32(3):300–313
Rawls J (1971) A theory of justice, rev. Harvard University Press, Cambridge, MA
Royal Decree (2021) Promulgating the basic statute of the state. Available at https://decree.om/2021/rd20210006/
Sherrington T (2017) The learning rainforest. John Catt Educational Ltd., Woodbridge:
Smith A (1759) The theory of moral sentiments. Indianapolis: Liberty Classics
The United Arab Emirates' Government portal (2023) We the UAE 2031' vision. Available at https://u.ae/en/about-the-uae/strategies-initiatives-and-awards/strategies-plans-and-visions/innovation-and-future-shaping/we-the-uae-2031-vision
Ullah AA, Lee SCW, Hassan NH, Nawaz F (2020) Xenophobia in the GCC countries: migrants' desire and distress. Global Affairs 6(2):203–223
United Nations (2016) FEATURE: UN uses virtual reality to inspire humanitarian empathy. Available at https://news.un.org/en/story/2016/05/529752-feature-un-uses-virtual-reality-inspire-humanitarian-empathy. Accessed 20th March 2024
United Nations (2023) The 17 goals. Available at https://sdgs.un.org/goals
UNSDG (2024) Universal values principle two: leave no one behind. Available at https://unsdg.un.org/2030-agenda/universal-values/leave-no-one-behind. Accessed 10th February 2024
Verkuyten M, Yogeeswaran K (2017) The social psychology of intergroup toleration: a roadmap for theory and research. Pers Soc Psychol Rev 21(1):72–96
Vescio V, Ross D, Adams A (2008) A review of research on the impact of professional learning communities on teaching practice and student learning. Teach Teach Educ 24(1):80–91
Oman Vision 2024 https://www.oman2040.om/vision-en.html
Vision 2030 (2023) A sustainable Saudi vision. https://www.vision2030.gov.sa/v2030/a-sustainable-saudi-vision/
Wren ST (2021) Improving Emirati students' social responsibility competence through global citizenship education. Learn Teach High Educ Gulf Perspect 17(2):80–94
Zaidan E, Al-Saidi M, Hammad SH (2019) Sustainable development in the Arab world–is the Gulf Cooperation Council (GCC) region fit for the challenge? Dev Pract 29(5):670–681
Zguir MF, Dubis S, Koç M (2021) Embedding Education for Sustainable Development (ESD) and SDGs values in curriculum: a comparative review on Qatar, Singapore and New Zealand. J Clean Prod 319:128534

Wendy A. Booth is a Lecturer in Public Service Management and Course Leader for Sociology at the University of South Wales. Wendy's specialist teaching areas relate to inequalities and include citizenship education and understanding community cohesion through the lens of tolerance, along with promoting ethnocultural empathy and mutual understanding using social psychology and sociological theories. Her research interests are, firstly, tolerance and ethnocultural empathy in relation to education, pedagogy and government policy, and using emerging technologies to support this. Secondly, the effectiveness of Artificial Intelligence in supporting mental health within conservative communities. Wendy's most recent book is titled Educating for Community Cohesion: Promoting Tolerance and Ethnocultural Empathy. She is the Managing Director of Sustainability Sphere - a Community Interest Company which focuses on teaching sustainability in schools, and TEEMA consultancy and research.

Part II

Institutionalizing Higher Education for Sustainable Development: Institutional and National Policies

8 Enhancing Climate Action Through Higher Education: The Role of GCC Universities in Advancing Nationally Determined Contributions (NDCs) and SDG 13

Quosay A. Ahmed, Tarig Z. Ahmed, Mawahib E. Ahmed, and Asia Mohamed

Abstract

This chapter examines the role of higher education institutions (HEIs) in the Gulf Cooperation Council (GCC) countries in advancing climate action and supporting national climate goals through specialized programs and research. Given the GCC's reliance on fossil fuels and its vulnerability to climate change, universities are strategically positioned to drive environmental awareness, build expertise, and foster innovation in sustainability. An analysis of climate-related programs across leading universities in Bahrain, Kuwait, Oman, Qatar, Saudi Arabia, and the UAE highlights disparities in program availability and diversity. While the UAE and Qatar lead the region in comprehensive climate education offerings, countries like Kuwait and Bahrain demonstrate significant gaps, particularly at the undergraduate level. This study identifies key areas for improvement, including the need for interdisciplinary and foundational programs across the GCC, and proposes collaborative strategies to enhance the region's climate education landscape. The chapter underscores the importance of HEIs in equipping the future workforce with the skills necessary to address the GCC's unique environmental challenges and achieve Sustainable Development Goals.

Keywords

Gulf Cooperation Council (GCC) · Higher education institutions · Climate education · Sustainable development · Renewable energy · Environmental science · Interdisciplinary programs · Nationally Determined Contributions (NDCs)

8.1 Introduction

The Gulf Cooperation Council (GCC) countries—comprising Bahrain, Kuwait, Oman, Qatar, Saudi Arabia, and the United Arab Emirates (UAE)—stand at a transformative crossroads as they navigate the dual challenges of economic growth and environmental sustainability. Traditionally reliant on hydrocarbon-based economies, GCC countries contribute significantly to global greenhouse gas (GHG) emissions and are among the highest CO_2 emitters per capita (European Commission 2021). Simultaneously, these nations are increasingly

Q. A. Ahmed (✉) · A. Mohamed
University of Khartoum, Khartoum, Sudan
e-mail: quosay.awad@uofk.edu

T. Z. Ahmed
University of Cambridge, Cambridge, UK

M. E. Ahmed
Institute of Engineering Research and Material Technology, National Center for Research of Sudan, Khartoum, Sudan

K. Al'Abri and E. Buckner (eds.), *Higher Education and Sustainable Development in the Arab Gulf States*, Sustainable Development Goals Series, https://doi.org/10.1007/978-3-032-19559-3_8

vulnerable to climate change impacts, including extreme heat, water scarcity, and rising sea levels (Our World in Data 2021). In response to these challenges, the GCC countries have committed to the Paris Agreement, advancing Sustainable Development Goals (SDGs), particularly SDG 13, which emphasizes Climate Action (United Nations n.d.).

One of the central components of the Paris Agreement is the concept of Nationally Determined Contributions (NDCs). NDCs represent each country's self-determined commitments to mitigate climate impacts by reducing emissions and building resilience through adaptation strategies (UN n.d.). As a testament to their dedication, GCC countries have articulated specific NDC targets aimed at reducing carbon emissions, diversifying energy sources, and enhancing climate resilience. These targets are not only integral to the Paris Agreement's global objectives but also resonate with national aspirations to establish resilient, diversified economies and sustainable societies. In line with this, Saudi Arabia, for example, aims to reduce its emissions by 278 million tons of CO_2 equivalent annually by 2030, focusing on renewable energy and improved energy efficiency (IRENA 2019).

Higher education institutions (HEIs) in the GCC region are uniquely positioned to support these national climate objectives. Universities and research institutions serve as incubators of knowledge and innovation, critical for advancing climate science, developing sustainable technologies, and promoting climate literacy. By integrating climate-focused education within degree programs, conducting impactful research, and fostering partnerships with governmental and private sectors, GCC universities play a vital role in translating NDC commitments into actionable, localized strategies. Notable institutions, such as the Masdar Institute of Science and Technology in the UAE and the Center for Sustainable Development at Qatar University, have already made strides in renewable energy research and environmental management, setting an example for other universities in the region (Masdar Institute 2023; Qatar University 2023).

However, the landscape of climate change education across the GCC is varied. A survey of 51 universities in the region reveals a diverse commitment to climate action in curricula, with UAE institutions generally taking the lead (QS Top Universities 2023). Some GCC universities offer specialized programs directly aligned with climate and sustainability topics, yet gaps persist, especially in integrating these programs at the undergraduate level. This disparity highlights the need for a coordinated approach across GCC countries to develop climate education that supports both the Paris Agreement and broader sustainability goals.

This chapter aims to explore how GCC higher education systems can drive transformative change in achieving NDCs. It will analyze the current climate education programs across universities in the region, identify the contributions and challenges of these institutions, and offer recommendations for strengthening the role of GCC universities in climate action. By mapping current efforts, this study seeks to inspire future initiatives and advocate for collaborative, future-focused planning within the Gulf region's higher education sector, positioning GCC universities as pivotal players in regional and global climate efforts.

8.2 Literature Review

8.2.1 Climate Change Context in the GCC

The Gulf Cooperation Council (GCC) countries occupy a significant position in global climate change discussions due to their large carbon footprints and reliance on hydrocarbon-driven economies. While the GCC collectively holds over 30% of the world's proven oil reserves and 23.6% of its gas reserves, the environmental costs of this resource dependency are substantial. Per capita, these countries have some of the highest greenhouse gas (GHG) emissions globally, driven by energy-intensive industries and high levels of energy consumption (European Commission 2021; Our World in Data

2021). Furthermore, the region faces environmental vulnerabilities that include rising temperatures, water scarcity, and ecosystem degradation, which underscore the urgency of transitioning toward sustainability.

Recognizing these challenges, GCC countries have aligned themselves with the Paris Agreement and are actively pursuing Sustainable Development Goals (SDGs), particularly SDG 13: Climate Action. Nationally Determined Contributions (NDCs) are integral to the Paris Agreement, representing each country's commitment to reduce GHG emissions and strengthen climate resilience. While NDCs vary across the GCC, the overarching targets include investing in renewable energy, enhancing energy efficiency, and fostering climate adaptation strategies (UN, n.d.). For example, and as shown is (Table 8.1), Saudi Arabia has committed to reducing emissions by 278 million tons of CO_2 equivalent annually by 2030, emphasizing renewable energy and efficiency improvements (IRENA 2019). These ambitious commitments position GCC countries as potential leaders in regional climate action. Table 8.1 summarizes the NDC targets and recent climate-related initiatives across GCC countries.

8.2.2 Role of Higher Education in Climate Change Education

Higher education institutions (HEIs) are recognized as crucial contributors to climate action through education, research, and policy support. Research indicates that universities can drive transformative change by promoting climate literacy, supporting scientific research, and equipping future leaders with the knowledge needed to tackle environmental issues (Bowers 2008; Lemons 2011). Scholars argue that HEIs must actively reform curricula to incorporate sustainability and climate change themes, ensuring that students not only understand the science but are also motivated to lead and innovate in sustainable development fields (Bowers 2008; Liu et al. 2015).

In recent years, climate-focused curricula and specialized sustainability programs have expanded globally. However, studies reveal that despite these advancements, many universities are still limited in their offerings. Shields (2019) suggests that while international student exchanges and environmental courses contribute positively to climate action, they also highlight the need for HEIs to mitigate their own environmental footprints. This point is particularly relevant in the GCC context, where high consumption and emission rates suggest that a "lead by example" approach within campuses could significantly impact the region's climate goals.

8.2.3 Climate Change Education in GCC Higher Education Systems

The landscape of climate change education within the GCC's HEIs is evolving but remains varied. Several leading institutions, such as the Masdar

Table 8.1 NDC targets and progress in GCC countries

Country	NDC target (by 2030)	Recent initiatives
Saudi Arabia	Reduce emissions by 278 million tons of CO_{2e}	Expansion in solar and wind projects
UAE	Increase solar energy capacity by 50%	Abu Dhabi solar farms, waste-to-energy projects
Qatar	Reduce GHG emissions by 25%	Investment in carbon capture and storage
Oman	Enhance water management, reduce GHG emissions	Water conservation and solar energy projects
Bahrain	Achieve 10% renewable energy in power generation	Solar rooftops in government buildings, energy efficiency
Kuwait	Reduce emissions intensity by 7.4%	Development of solar farms and clean energy infrastructure

Sources: (IRENA, 2023) and UNFCCC

Table 8.2 Climate-related programs in leading GCC universities

University	Country	Program title	Degree level	Focus area	Funding source
Masdar Institute	UAE	Renewable energy and climate mitigation	Postgraduate	Renewable energy, climate adaptation	Government, private partnerships
Qatar University	Qatar	Environmental science	Undergraduate	Environmental management	Government, industry partnerships
King Abdullah University of Science and Technology (KAUST)	Saudi Arabia	Environmental Science and Engineering (EnSE)	Postgraduate	Environmental policy, resilience and Engineering	Government
University of Bahrain	Bahrain	Environment and sustainable development	Postgraduate	Sustainability, resource management	Government
Sultan Qaboos University	Oman	Environmental Science	Postgraduate	Sustainable urban development	Government, international partnerships
Prince Sultan University	Saudi Arabia	Sustainable business practices	Undergraduate	Business sustainability	Private sector

Sources Masdar Institute of Science and Technology (2023). Campus Programs; Qatar University (2023). Center for Sustainable Development; King Abdullah University of Science and Technology (KAUST, 2024). Environmental Science and Engineering (EnSE); University of Bahrain (2024). Master in Environment and Sustainable Development; Sultan Qaboos University (2023/2024). MSc in Environmental Science; Prince Sultan University (2023). Sustainable Development Courses

Institute of Science and Technology in the UAE and the Center for Sustainable Development at Qatar University, are recognized for their specialized climate and sustainability programs (Masdar Institute 2023; Qatar University 2023). These institutions offer courses that span renewable energy, environmental science, and sustainable development, aligning their educational goals with national climate commitments.

Nevertheless, a broader survey of 51 universities across the GCC reveals considerable gaps in climate-focused curricula, particularly at the undergraduate level (QS Top Universities 2023). While UAE-based universities lead in offering specialized sustainability programs, many universities in Bahrain, Kuwait, and Saudi Arabia lack comprehensive degree programs focused on climate action and sustainable development. This disparity reflects broader challenges within the GCC's HE systems, where traditional reliance on hydrocarbon industries and a lack of institutional emphasis on environmental studies may limit the advancement of climate education. Table 8.2 presents selected climate-related programmes in leading GCC universities.

8.2.4 Higher Education's Impact on Nationally Determined Contributions (NDCs)

Research underscores the role of HEIs in advancing NDC targets by fostering climate research, building climate resilience, and creating a skilled workforce to support sustainable economies. By providing students with the knowledge and skills needed to address environmental issues, HEIs contribute directly to achieving NDC targets. For instance, partnerships between universities and government bodies can support policy formation, while industry collaborations help develop and implement renewable energy technologies (IRENA 2019). Table 8.3 shows the

Table 8.3 Renewable energy capacity growth in GCC (2019–2023), after IRENA (2023)

Country	Renewable capacity in 2019 (GW)	Renewable capacity in 2023 (GW)	Growth (%)
UAE	5.2	13.6	161
Saudi Arabia	1.2	10.5	775
Qatar	0.5	2.5	400
Oman	0.7	3.6	414
Bahrain	0.1	0.6	500
Kuwait	0.2	1.6	700

growth in renewable energy development in GCC countries during 2019–2023.

In the GCC, universities are beginning to play more proactive roles in NDC-aligned research. Studies suggest that collaborations between GCC universities and governmental agencies can enhance capacity-building efforts, particularly in sectors like renewable energy, water management, and urban planning (Al-Fanar Media 2019). However, many universities lack resources for comprehensive NDC-supporting initiatives, which limits their ability to contribute effectively to climate action. Expanding funding for climate research and embedding climate science in curricula across more universities would strengthen the overall contribution of GCC HEIs to national climate goals.

8.2.5 Gaps and Opportunities in GCC Climate Change Education

Despite notable advancements, several gaps remain in the GCC's climate education landscape. A lack of centralized resources and data hampers understanding of the breadth and depth of climate change education programs across the GCC. While the UAE leads in climate-focused programs, other countries display limited offerings. According to UNESCO (2019), expanding climate education in the GCC is essential for meeting regional climate goals and supporting a long-term shift toward sustainability.

Research also indicates the need for GCC universities to incorporate "green campus" initiatives to model sustainable practices directly. By adopting renewable energy sources, promoting energy efficiency, and reducing waste, universities can demonstrate a practical commitment to climate action and inspire similar initiatives across other sectors (Shields 2019). Further, collaboration with international HEIs could bolster GCC universities' capacity to adopt best practices and innovative models in climate education.

8.3 Research Objectives and Methods

8.3.1 Research Objectives

This study explores how higher education institutions (HEIs) in the Gulf Cooperation Council (GCC) contribute to Nationally Determined Contributions (NDCs) and Sustainable Development Goal (SDG) 13: Climate Action. The objectives are to:

1. **Map Climate Education Programs**: Identify climate-related programs in top GCC universities, categorizing them by level (undergraduate/postgraduate) and focus (e.g., sustainable development, Environmental Engineering).
2. **Evaluate HEIs' Impact on NDCs**: Assess how educational and research programs support national climate goals, especially in renewable energy, environmental management, and resilience (UN, n.d.).
3. **Identify Gaps and Propose Strategies**: Highlight limitations in current offerings and recommend improvements for expanding climate education across the region.

8.3.2 Methodology

The study uses secondary data from official university websites, educational portals, and institutional reports to provide a comprehensive view of climate education in GCC HEIs. This multi-source approach is designed to mitigate potential data gaps from any single source. Key aspects include:

- **University Selection**: Fifty-one leading universities were selected based on regional rankings (QS Top Universities 2023), ensuring representation across Bahrain, Kuwait, Oman, Qatar, Saudi Arabia, and the UAE.
- **Data Collection**: Climate-related programs were identified and categorized by level and specialization. Keywords such as "Climate Change," "Sustainable Development," and "Environmental Science" guided the search across university websites and the Study Portals database (Study Portals 2023).
- **Limitations**: Online data may be incomplete, as some program offerings are not fully disclosed. Excluding smaller specialized institutions may also overlook unique programs (Al-Fanar Media 2019).

Furthermore, only institutes with bachelors or master's degree courses are considered. PhD courses were excluded from this research. There are several small specialists institutes and colleges in the GCC; however only large multidisciplinary universities with several faculties are considered. This is to ensure the data is non-biased and comparable. Since small institutes with specialties outside of science, physics, and engineering are unlikely to have climate change related programs.

This method provides an overview of climate education in GCC universities, while recognizing areas for further improvements, such as:

1. Triangulation: In future studies, it is better to contact university representatives directly to verify information and potentially uncover programs not clearly listed on websites.
2. Broader Keyword Search: Include synonyms or related terms to capture a wider range of programs but carefully review results to ensure they align with the research focus.
3. Consider a Pilot Study: Test the methodology on a small sample of universities to identify any potential issues before large-scale data collection.

8.4 Data and Findings

8.4.1 Climate-Related Programs Across Universities

The analysis as illustrated in Table 8.4 shows varied climate education offerings among GCC universities. UAE institutions lead, with 8 out of 10 surveyed universities offering degrees in sustainable development, environmental science, or renewable energy, notably through institutions like the Masdar Institute and Khalifa University (Masdar Institute 2023). In Qatar, institutions such as Qatar University and Hamad Bin Khalifa University actively support climate education through research centers focused on sustainable development (Qatar University 2023).

In contrast, Bahrain and Kuwait offer fewer climate-focused programs, with only one out of the five top universities in Kuwait offering an environmental science program. This uneven distribution highlights disparities in climate education, suggesting room for growth in these areas (UNESCO 2019).

The following table presents a summary of climate change, sustainable development, and environment-related programs at select top-tier universities across the GCC region (51 Universities), identified through analysis of official university websites and the Study Portals website. It includes both Master's (MSc or MEng) and Bachelor's (BSc or BEng) degrees, where available.

8.4.2 Alignment with National Climate Goals

Universities focus their climate programs in three main areas:

1. **Renewable Energy and Sustainable Development**: Leading universities in the UAE and Qatar offer programs aligned with national

Table 8.4 List of top international universities in the GCC and their available programs related to climate change, sustainable development and the environment

Country	Universities	Masters (MSc or MEng)	Bachelor (BSc or BEng degree)
Bahrain	1. Applied science university—Bahrain	None	None
	2. Ahlia University	None	None
	3. University of Bahrain	1. Environment and sustainable development	1. Renewable energy engineering
	4. Arabian Gulf University	1. Environmental management	None
		2. Environmental sciences and natural resource	
		3. Desert cultivation technologies and soilless agriculture	
	5. The Kingdom University	None	1. Civil and environmental engineering
Kuwait	1. American University of the Middle East (AUM)	None	None
	2. Kuwait University (KU)	None	1. Environmental sciences
	3. Gulf University of Science and Technology (GUST)	None	None
	4. Arab Open University (AOU)—Kuwait	None	None
	5. Kuwait College of Science and Technology	None	None
Oman	1. Sultan Qaboos University	1. Environmental science	None
	2. University of Nizwa	None	1. Environmental engineering
	3. National University of Science and Technology	None	None
	4. Dhofar University	None	1. Civil and environmental engineering
	5. German University of Technology in Oman (GUTECH)	None	1. Environmental engineering
	6. Sohar University	1. Environmental engineering	None
Qatar	1. Qatar University	1. Environmental engineering	None
		2. Environmental science	
	2. Hamad Bin Khalifa University (HBKU)	1. Sustainable environment	None
		2. Sustainable energy	
	3. Texas A&M University at Qatar (TAMUQ)	None	None
	4. Northwestern University in Qatar (NU-Q)	None	None
	5. Lusail University	None	None
Saudi Arabia	1. King Abdulaziz University	1. Crisis management in the environment	1. Civil engineering and environment
		2. Civil engineering and environment	2. Environmental sciences
		3. Environmental sciences	

(continued)

Table 8.4 (continued)

Country	Universities	Masters (MSc or MEng)	Bachelor (BSc or BEng degree)
	2. King Fahd University of Petroleum and Minerals	1. Environmental sciences	1. Civil and environmental engineering
	3. King Saud University	None	None
	4. Umm Al-Qura University	None	None
	5. Prince Mohammad Bin Fahd University	None	None
	6. Princess Nourah bint Abdulrahman	None	None
	7. Imam Abdulrahman Bin Faisal University	1. Environmental health 2. Environmental engineering	1. Environmental engineering
	8. King Faisal University	None	1. Civil and environmental engineering
	9. King Khalid University	None	None
	10. Alfaisal University	None	None
	11. Imam Mohammad Ibn Saud Islamic University—IMSIU	None	None
	12. Jouf University	None	None
	13. Prince Sultan University	None	1. Civil and environmental engineering
	14. AlMaarefa University	None	None
	15. Effat University	None	None
	16. Northern Borders University	None	None
	17. Qassim University	None	None
	18. Prince Sattam Bin Abdulaziz University	None	None
	19. Taibah University	None	None
	20. Jazan University	None	None
UAE	1. United Arab Emirates University (UAEU)	1. Sustainable critical infrastructure 2. Health, safety and environmental engineering	None
	2. Khalifa University	1. Sustainable critical infrastructure 2. Water and environmental engineering 3. Health, safety and environmental engineering	None
	3. American University of Sharjah (AUS)	None	1. Environmental science 2. Environmental and water engineering 3. Environmental policy
	4. University of Sharjah	1. Sustainable and renewable energy	1. Environmental science and engineering 2. Sustainable and renewable energy

(continued)

Table 8.4 (continued)

Country	Universities	Masters (MSc or MEng)	Bachelor (BSc or BEng degree)
	5. Zayed University	1. Sustainability	1. Environment and sustainability sciences
	6. Canadian University Dubai	1. Public health—environmental health management	None
	7. Ajman University	None	None
	8. Al Ain University	None	None
	9. Abu Dhabi University	None	1. Environmental health and safety
	10. American University in Dubai	None	1. Environmental engineering (Minor)

goals for renewable energy and sustainability, meeting targets for cleaner energy and improved efficiency (IRENA 2019).

2. **Environmental Engineering and Management**: Programs in Environmental Engineering, available in institutions like Qatar University and KAUST in Saudi Arabia, support adaptation strategies for water conservation, waste management, and sustainable urban planning (Al-Fanar Media 2019).
3. **Climate Resilience and Adaptation**: Some institutions, such as Qatar University's Center for Sustainable Development, emphasize resilience, though programs dedicated to adaptation remain limited across the region.

Based on the information provided in Table 8.4, the following table summarizes the status of climate-related educational efforts in GCC countries and showing gaps, implications and insights for each country (Table 8.5).

Nonetheless, several gaps remain in universities' contributions to climate action:

- **Limited Undergraduate Offerings**: Climate education predominantly exists at the postgraduate level, limiting exposure for undergraduate students. Incorporating climate and sustainability topics into undergraduate curricula across more disciplines would provide foundational knowledge early in students' education (UNESCO 2019).
- **Narrow Disciplinary Focus**: Few business, humanities, or social science programs include climate content, which is vital for preparing a workforce ready to address sustainability challenges across sectors (Shields 2019).
- **Absence of a Centralized Resource**: Without a shared resource for climate-related programs, institutions lack coordination, and students face barriers in accessing program information. Establishing a regional climate education platform could enhance collaboration and support joint research efforts.

8.5 Discussion and Analysis

8.5.1 Contribution of GCC HEIs to Climate Goals

GCC universities show potential to support national climate goals through targeted educational programs and research centers. As demonstrated by institutions like the Masdar Institute in the UAE and Qatar University's Center for Sustainable Development, certain universities have taken steps to align their programs with NDC targets by focusing on renewable energy, sustainable infrastructure, and climate resilience (Masdar Institute 2023; Qatar University 2023). However, a regional disparity exists, with universities in Bahrain, Kuwait, and, to some extent, Saudi Arabia lagging in climate-

Table 8.5 GCC climate education landscape: current status, gaps, recommendations, and strategic insights by country

Country	Current status	Key gaps	Implications/ recommendations	Insights
Bahrain	Limited climate-related programs with only a few offerings at University of Bahrain and Arabian Gulf University. Focus areas include environment and sustainable development and desert cultivation	Minimal undergraduate and graduate options; few programs in renewable energy and broad climate science topics	Expand programs in renewable energy, climate policy, and interdisciplinary environmental studies to meet local needs such as water scarcity and resilience	Bahrain's geographical challenges (e.g., water scarcity, desertification) make it uniquely suited for programs in water management and desert agriculture
Kuwait	Fewest climate-focused offerings among GCC countries. Only one bachelor's program in environmental sciences at Kuwait University, with no master's programs	No graduate programs and limited undergraduate options across universities, despite high per capita emissions	Develop foundational programs at both levels, especially in environmental science, energy efficiency, and climate mitigation, to align with climate goals	Kuwait's reliance on fossil fuels provides an opportunity to lead in transitioning toward renewable energy, with potential focus on sustainable energy education
Oman	Moderate number of programs, primarily engineering-focused (e.g., environmental science at Sultan Qaboos University, Environmental Engineering at Sohar University)	Limited diversity in fields; few programs in renewable energy and sustainability management	Introduce broader programs in climate adaptation, renewable energy, and environmental management to support resilience and urban planning goals	Oman's focus on sustainability in tourism and urban planning can be supported by specialized programs in sustainable urban development and eco-tourism
Qatar	Advanced graduate programs at Qatar University and HBKU, including sustainable environment and sustainable energy, but lacks undergraduate climate programs	Absence of undergraduate climate-focused programs restricts entry-level education and foundational skill-building	Expand undergraduate programs in climate science, environmental engineering, and sustainable development to prepare a skilled workforce for national goals	Qatar's leadership in clean energy and sustainability can be enhanced through greater undergraduate engagement, building a pipeline of local climate experts
Saudi Arabia	Largest number of universities but limited climate-related offerings. Some specialized programs at King Abdulaziz University and King Fahd University (e.g., Crisis management in environment)	Underrepresentation of climate science, environmental policy, and renewable energy programs, especially at the master's level	Develop interdisciplinary programs in climate adaptation, environmental health, and sustainable energy to support vision 2030 and build relevant expertise	Saudi Arabia's vision 2030 goals present an opportunity to scale climate education, particularly in fields supporting national infrastructure and policy
UAE	Most comprehensive climate-related education landscape in the GCC. Universities like UAEU and Khalifa University offer diverse programs at both undergraduate and graduate levels	Few gaps; however, there is potential to expand interdisciplinary programs combining policy, engineering, and environmental sciences	Strengthen interdisciplinary programs and promote regional partnerships to share curriculum and expertise with other GCC countries	The UAE's established position as a leader in renewable energy and sustainability in the GCC makes it an ideal partner for regional education initiatives

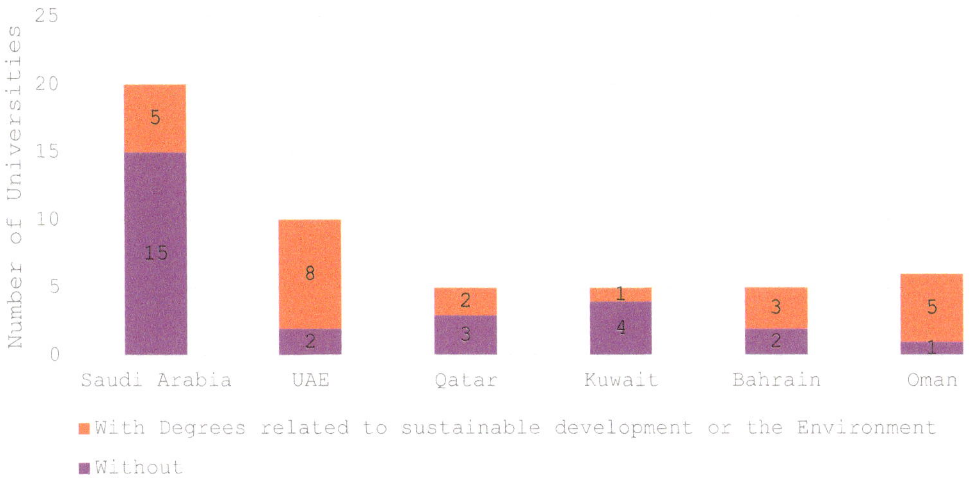

Fig. 8.1 Universities with and without degrees related to sustainable development and climate change in the GCC

oriented academic offerings (UNESCO 2019). This uneven progress indicates a need for more uniform commitment across GCC universities.

The importance of higher education in supporting climate resilience is further highlighted by studies showing that HEIs play a critical role in fostering awareness and developing a skilled workforce for sustainable development (Bowers 2008; Lemons 2011). While several GCC universities have adopted climate-related postgraduate programs, these initiatives remain limited at the undergraduate level, which restricts early exposure to climate science and sustainability topics. Integrating these themes in undergraduate studies would broaden foundational climate knowledge and prepare a greater number of students to contribute to sustainability efforts in their future careers.

Figure 8.1 shows the GCC countries with and without degrees in UN SDG 13 climate action and sustainability related topics. Out of all the GCC countries studied, the UAE had one of the highest rates of courses in sustainability and the environment; 8 out of the 10 universities had some of these courses. Some of the courses were newly added. For example, Zayed university introduced an MSc degree in Environmental and Sustainability Sciences as recently as April 2023 (Zayed University 2023). Some universities even have dedicated pages on their website focusing on their compliance with the UN SDG (Abu Dhabi University 2023). Even for the two universities with no specific course on sustainable development or the environment there was still a heavy emphasis on sustainability within its other courses. A good example is the Bachelor of Architectural Engineering course at Ajman University (Ajman University 2023). That university also has a dedicated page on sustainability.

Furthermore, out of the 51 universities investigated across the GCC 24 universities were found to have degrees related to the UN SDG 13 topic of Climate Action. Such as sustainable development, Environmental Engineering, or crisis management in the environment. On closer analysis, the majority of these universities had either a masters degree program on the subjects or a bachelors. Only six out of these 24 universities had both a Bachelor and Master's degree related to these subjects (Fig. 8.2).

8.5.2 Addressing Gaps and Enhancing Regional Collaboration

One of the primary findings of this study is the lack of interdisciplinary integration of climate topics across various academic disciplines, such as business and social sciences. Shields (2019)

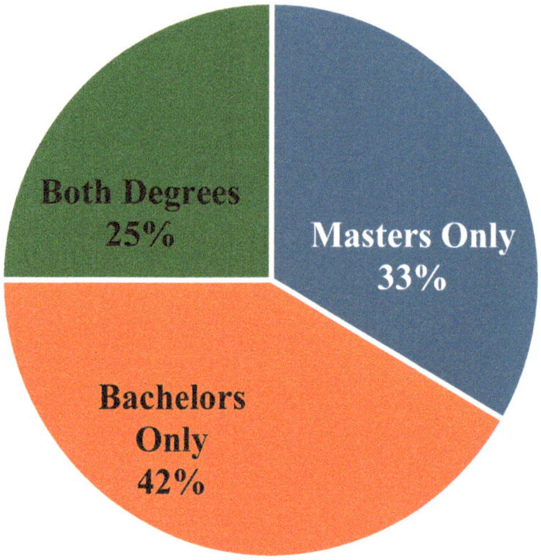

Fig. 8.2 Comparison of Universities with degrees related to sustainable development and climate change in the GCC and their type (24 out of 51 universities)

emphasizes that a well-rounded approach, incorporating climate and sustainability education across disciplines, is essential for a comprehensive understanding of climate impacts and solutions. Expanding climate content to diverse fields could cultivate a workforce capable of applying sustainability principles in multiple sectors, such as policy, finance, and infrastructure.

The absence of a centralized platform for climate education in the GCC poses another challenge, as this limits knowledge-sharing and collaborative opportunities among universities. Establishing a GCC-wide resource or network could facilitate the exchange of best practices, joint research projects, and climate-focused educational strategies. Models from global consortia, such as the United Nations University, could serve as examples for creating a regional body that connects GCC HEIs in addressing climate action through higher education (UNESCO 2019).

8.5.3 Green Campus Initiatives

GCC universities also have an opportunity to adopt green campus initiatives, which would demonstrate climate action in practice. By integrating renewable energy sources, energy-efficient technologies, and waste reduction programs on campuses, universities could reduce their own carbon footprints, setting an example for other institutions and sectors. This "lead by example" approach aligns with the region's emphasis on sustainable development as outlined in NDCs, particularly in energy-intensive areas where universities can showcase feasible solutions (IRENA 2019).

Implementing green initiatives within campuses could also serve as educational tools, allowing students to learn firsthand about sustainable practices and their impacts. This approach is particularly relevant in the GCC, where per capita energy consumption and emissions are among the highest globally (European Commission 2021). Adopting sustainable operations within HEIs could help bridge the gap between climate theory and practice, fostering a more sustainability-conscious community.

8.6 Conclusions and Implications

This study has demonstrated the important role of GCC universities in supporting climate action, identifying both achievements and areas for growth. While the UAE and Qatar have made significant progress in offering climate-related programs, universities in Bahrain and Kuwait still face challenges in integrating sustainability education comprehensively. The analysis highlights the alignment of certain educational programs with NDC targets in areas like renewable energy, Environmental Engineering, and sustainable development, underscoring the potential of GCC universities to advance national and regional climate goals.

However, key gaps persist, including limited interdisciplinary climate education and a lack of undergraduate programs that incorporate sustainability. Additionally, the absence of a regional climate education network hinders collaboration and knowledge exchange across GCC HEIs, suggesting an opportunity for a coordinated approach to climate action in higher education.

To strengthen the role of GCC universities in climate action, the following recommendations are proposed:

1. **Expand Undergraduate Climate Programs**: Introduce climate and sustainability topics across more undergraduate disciplines, ensuring that students gain foundational knowledge early in their education. Such exposure would better prepare future professionals to contribute to climate resilience in diverse sectors.
2. **Promote Interdisciplinary Climate Education**: Encourage integration of climate education into business, social sciences, and other fields to cultivate a holistic understanding of sustainability challenges and solutions. This expansion aligns with findings by Shields (2019), who argues for a cross-disciplinary approach to climate education.
3. **Establish a Regional Climate Education Network**: A centralized GCC climate education network would enhance collaboration, facilitating resource-sharing, joint research, and best practices for climate-focused academic programs (UNESCO 2019).
4. **Research and Development (R&D)**: The universities are recommended to conduct R&D activities to identify ways to reduce greenhouse gas emissions and adapt to climate change. This includes developing and/or customizing abatement technologies for GHG emissions, promoting renewable energy sources (solar, wind and geothermal), and improving energy efficiency.
5. **Implement Green Campus Initiatives**: HEIs should adopt renewable energy, waste reduction, and energy-efficient technologies on their campuses, promoting sustainable practices that align with NDC goals. These initiatives would position GCC universities as role models in regional climate action and offer experiential learning opportunities for students.

Future studies could explore the impact of current GCC climate education programs on students' professional outcomes, particularly in sectors critical to national climate goals. Further research could also investigate the feasibility of a GCC-wide green campus framework, measuring the potential environmental benefits and educational impacts of such initiatives.

By addressing current limitations and leveraging collaborative opportunities, GCC universities can enhance their role in climate action, contributing to both regional sustainability and the global response to climate change.

References

Abu Dhabi University (2023) Sustainable development goals. Sustainable development goals. Retrieved from https://www.adu.ac.ae

Ajman University (2023) Bachelor of architectural engineering. Bachelor of architectural engineering. Retrieved from https://www.ajman.ac.ae

Al-Fanar Media (2019) Gulf countries invest heavily in higher education but challenges remain. Retrieved from https://www.al-fanarmedia.org

Bowers CA (2008) University reform in an era of global warming. Ecojustice Press

European Commission (2021) GHG emissions of all world—publications Office of the EU. https://op.europa.eu/en/publication-detail/-/publication/45c88a84-2d65-11ec-bd8e-01aa75ed71a1/language-en

International Renewable Energy Agency (IRENA) (2019) Renewable energy market analysis: GCC 2019. https://www.irena.org/publications/2019/jan/renewable-energy-market-analysis-gcc-2019

IRENA. Renewable energy markets: GCC 2023, *International Renewable Energy Agency (IRENA)*, ISBN: 978-92-9260-568-1 (2023).

King Abdullah University of Science and Technology (KAUST) (2024) Environmental Science and Engineering (EnSE), academic programme page. Thuwal: KAUST.

Lemons J (2011) The urgent need for universities to comprehensively address global climate change across disciplines and programs. Environ Manage 48(3):379–391

Liu S, Roehrig G, Varna K (2015) In-service teachers attitudes, knowledge, and classroom teaching of global climate change. Sci Educ 24(1)

Masdar Institute of Science and Technology (2023) [Institutional website]

Our World in Data (2021) Per capita electricity generation. Retrieved from https://ourworldindata.org/grapher/per-capita-electricity-generation?tab=table

Prince Sultan University (2023) Sustainable development goals: PSU's commitment to SDG 2030. Retrieved from https://www.psu.edu.sa

Qatar University (2023) Centre for sustainable development. [Institutional website]

QS Top Universities (2023) QS arab region university ranking. Retrieved from https://www.topuniversities.com

Shields R (2019) The sustainability of international education: student mobility and global climate change. J Clean Prod 217:594–602

Study Portals Masters (2023) Masters degrees in environmental science. Retrieved from https://www.mastersportals.com

Sultan Qaboos University (2023/2024) MSc in Environmental Science, College of Science / Deanship of Postgraduate Studies. Muscat: Sultan Qaboos University.

UNESCO (2019) Education in the GCC countries: key trends and statistics. Retrieved from https://unesdoc.unesco.org

University of Bahrain (2024) Master in Environment and Sustainable Development, College of Science. Sakhir: University of Bahrain. (College of Science)

United Nations (n.d.) All about the NDCs. UN climate change. Retrieved from https://www.un.org/en/climatechange/all-about-ndcs

Zayed University (2023) Zayed university launches a master of science in environment and sustainability sciences. Retrieved from https://www.zu.ac.ae

Quosay Awad Ahmed PhD, is an Assistant Professor at the University of Khartoum. His research interests encompass energy, Greenhouse Gases (GHG) accounting, climate change mitigation, sustainability education, and capacity building needed to meet national climate objectives.

Tarig Z. Ahmed is a Research Associate at the University of Cambridge. His major research interests lie in sustainable engineering, environmental policy, and the integration of climate action within educational institutions.

Mawahib E. Ahmed PhD, is a Senior Research Scientist at the Institute of Engineering Research and Material Technology, National Center for Research of Sudan. Her research focuses on energy, biofuel, material engineering, environmental sustainability, and advancing technology for climate resilience.

Asia Mohamed PhD, is a Lecturer at the Institute of Environmental Studies in the University of Khartoum. Her primary research areas include environmental science, climate change adaptation, and socioeconomic aspects in the environmental studies.

Green Emissaries: The Climate Change Leadership of Branch Campuses in the Arab Gulf States

9

Seungah Sarah Lee and Gabriela Reynell

Abstract

International branch campuses in the Arab Gulf are uniquely positioned to contribute to global and national sustainability goals, as they navigate the dual embeddedness of the parent (home) institution's sustainability initiatives and the host country's environmental priorities. This chapter observes how two American branch campuses in the Arab Gulf present and engage in their sustainability efforts. We find that sustainability narratives across the branch campuses and their home institutions are fairly similar and attuned to global rhetoric. Simultaneously, however, we find that the branch campuses are attuned to the sustainability initiatives and goals of their host countries, whereby these campuses leverage their positions as "elite" international branch campuses to envision themselves as leaders and proactive participants of their host countries' sustainability efforts, thereby, in some ways, using these efforts as a way to buffer against critiques of these campuses not being as locally embedded or engaged.

Keywords

Sustainability · Organization theory · Embeddedness · Higher education · Arab Gulf · Branch campuses

9.1 Introduction

The Talloires Declaration of 1990 affirmed publicly for the first time the crucial role of higher education in the global sustainability movement (Sylvestre et al. 2013). Since then, an increasing number of universities worldwide have taken a vested interest in environmental sustainability, as reflected in emerging curricular offerings, research agendas, and efforts to mitigate climate change by incrementally greening campus operations. Although the Arab Gulf states have been relative latecomers in adopting policies to promote environmental sustainability and climate change action, the recent push for environmental conservation, sustainability, and net-zero initiatives of the Gulf states has led to an expectation that Gulf universities will assume an active, fundamental role in the sustainability movement. Universities in the Arab Gulf have, in fact, been increasingly branded as hubs for green technology for sustainability (Koch 2018).

Among the multitude of higher education institutions in the Arab Gulf, international branch campuses are in a unique position to contribute to global and national sustainability goals. On

S. S. Lee (✉)
HEC Paris, Doha, Qatar
e-mail: lees@hec.fr

G. Reynell
University of California, Berkeley, USA
e-mail: gabireynell@berkeley.edu

K. Al'Abri and E. Buckner (eds.), *Higher Education and Sustainable Development in the Arab Gulf States*, Sustainable Development Goals Series, https://doi.org/10.1007/978-3-032-19559-3_9

the one hand, branch campuses are physically and culturally embedded in their host countries (Shams and Huisman 2016). On the other hand, they maintain their parent unit's identity across borders, aligning its ethos, model, and curriculum to the home campus. In other words, branch campuses respond to dual embeddedness (Figueiredo 2011), having to pay simultaneous attention to what is happening across oceans in their home country and the requirements and demands of their host countries. Hence, when it comes to sustainability, it is likely that Gulf branch campuses pay attention to the sustainability initiatives and efforts of their parent campus, drawing from resources and learnings from their home country. Simultaneously, they likely respond to the host Gulf country's emergent vision, strategies, goals, and initiatives surrounding environmental sustainability and climate change action. Since parent campuses are often located in countries (e.g., the U.S.) where the sustainability movement came earlier and has had time to mature (Filho 2000), the branch campuses are likely to be attuned to sustainability goals and action plans of the parent institution, leaving these branch campuses to be better positioned to launch and carry out sustainable initiatives compared to their counterparts in the host country. Yet, although branch campuses inherit and can easily imitate the parent campus's sustainability initiatives, they must also be able to successfully translate them to the Arab Gulf context to achieve their full potential (Koch 2018; Sahlin and Wedlin 2008).

This chapter is guided by the following questions: How are branch university campuses in the Arab Gulf addressing sustainability and climate change action in response to their dual embeddedness? In other words, to what extent are the branch campuses' initiatives and efforts aligned with or informed by the sustainability orientation of their parent campuses? And how might their sustainability efforts be further aided or propelled by the emergent sustainability policies and priorities of their host countries?

To answer these questions, we qualitatively examine the online presence of branch campuses vis-à-vis sustainability as reflected in university websites. We also contextualize the sustainability orientation of branch campuses in the Arab Gulf states through official documents and related memoranda of understanding at the local, national, or regional level. We have selected two branch campuses as illustrative case studies for this chapter: New York University Abu Dhabi (NYUAD) and Georgetown University in Qatar (GUQ).

9.2 Literature Review

9.2.1 The Rise of a Global Higher Education Sustainability Regime

In this chapter, we follow the original definition codified by the United Nations in the mid-1980s through the Brundtland Report, whereby "sustainability" (or "sustainable development") is understood broadly as encompassing not just environmental, but also social and economic considerations, with the ultimate goal of meeting "the needs of the present without compromising the ability of future generations to meet their own needs" (WCED 1987). As the environmental regime expanded and intensified over the twentieth century and with universities increasingly acting as "receptor sites" for global ideas (Frank 2000), it did not take long for higher education to embrace sustainability in this very capacious definition. In fact, a wave of commitments and declarations have been made by universities everywhere, suggesting that sustainability is becoming inextricably linked to the higher education sector. Declarations, such as the Halifax Declaration of 1991, the University Charter for Sustainable Development of 1994, the Ubundtu Declaration of 2002, and the Turin Declaration on Education and Research for Sustainable Development of 2009, all stressed the important role universities can play in supporting and incorporating sustainability practices around the world. In 2019, networks representing more than 7,000 higher education institutions from six continents declared a "Climate Emergency," and over 1,000 of those individual universities subsequently signed up for the "Race to Zero," a global initiative for a carbon–

neutral future spearheaded by the higher education sector. Among the signatories of this latest initiative are more than 10 universities based in the UAE, three universities in Saudi Arabia, and one university in Qatar.

In addition to these symbolic commitments, the rise of the global higher education sustainability regime has been propped up in recent years by the emergence of a multitude of public rankings and awards meant to highlight the comparative performance of universities with respect to various sustainability metrics. For instance, the Times Higher Education (THE) started publishing its "Impact Rankings" in 2019, through which the sustainability performance of universities from around the world is assessed against the UN Sustainable Development Goals (SDGs). The ranking claims to provide "a comprehensive and balanced comparison" across four broad areas (research, stewardship, outreach, and teaching). In the U.S., similar rankings emerged even earlier. The Princeton Review has been publishing its popular "Guide to Green Colleges" since 2009; the Sierra Magazine has been publishing its own list of "Cool Schools" since 2007, with the top colleges and universities on that list awarded the coveted "Eco-valedictorian" title.

Despite these developments, some express skepticism over the actual impact of symbolic sustainability commitments made by universities, which are reminiscent of "greenwashing" concerns that have emerged in the public sector (Jones 2012; Disterheft et al. 2013). Observers of this phenomenon have also warned against the unintended consequences of publicly ranking universities based on their sustainability performance, which may skew their incentives for greening in the first place. Notwithstanding these concerns, the intensification of these trends, along with the recent professionalization of the field of higher education sustainability, has led to universities across the board paying increasingly more attention to sustainability in recent years, and a slew of organizational practices being adopted as a result.

One recent study conducted on a nationally representative sample of U.S. higher education institutions (Gavrila 2023) found that as many as 63% of the colleges and universities surveyed had set up a dedicated sustainability webpage (as of 2020), and 57% had established a sustainability office (or some other form of governing structure); interestingly, almost a third (31%) of those had signed onto a major sustainability pledge that publicly committed them to improving their sustainability performance. Additionally, many of these institutions had begun drafting internal strategic documents that documented the steps they would take to conduct carbon emission assessments on their campuses, or step-by-step Climate Action Plans (CAPs). It is worth noting that many of these institutions chose to express their commitments to sustainability in similar organizational ways, suggesting that strong mimetic, normative, and competitive forces at work within the higher education system as a whole, in keeping with the isomorphism hypothesis put forth by DiMaggio and Powell (1983), as well as a sense of ceremonial adoption of these practices that develops over time (Meyer and Rowan 1977). Whether the same isomorphic forces that prompt colleges and universities in the U.S. to adopt similar sustainability practices and structures also apply to branch campuses located elsewhere is a different question and an open one for that matter.

9.2.2 The Role of and Dual Embeddedness of Branch Universities in the Gulf States

The motivations for establishing and maintaining branch campuses are complex and varied. For the host countries, branch campuses bring closer the prestige of a foreign university along with new ideas about curriculum, governance, teaching, and other innovations and help raise the profile and image of the host country in the global arena as a form of "soft power" initiatives (Lee 2021; Altbach and de Wit 2020). Additionally, branch campuses have been a part of higher education reform efforts aimed at providing nationals in host countries with global knowledge and skills that will allow them to participate more productively in the labor market and contribute to their national economies.

Among the Gulf countries, Qatar and the UAE have both adopted progressive education models and imported globally recognized universities to develop their respective higher education systems as part of their internal education reforms. Qatar invited a number of select foreign universities (including Georgetown University, Carnegie Mellon University, Northwestern University, Texas A&M, Virginia Commonwealth University, HEC Paris, etc.) to establish branch campuses and offer specific, renowned academic programs in forming its Education City, officially inaugurated in 2003 (Ibnouf et al. 2014; Lee 2021). The Emirati case is more extensive than that of Qatar, with the UAE hosting over 40 international branch campuses across its different emirates, including New York University Abu Dhabi (NYUAD), which is a full-scale research university that replicates its main campus in New York City (Becker 2015; White 2015).

In particular, Qatar's American branch campuses in Education City can be seen as investments in a rising knowledge economy that primarily fosters nativist national identity and Qatarization, with the purposed goal of "unlocking the human potential" of Qataris themselves (Vora 2018). Likewise, the international branch campuses of the UAE, including NYU Abu Dhabi, are part of the UAE's vision and desire to become more economically competitive and develop local research capacities (Lane and Pohl 2017), which may give it an advantage in the region. Largely funded by the Qatari and Emirati government, high-profile foreign universities such as the American branch campuses in Education City, Qatar, and NYUAD thus play an important role in the national and regional higher education systems.

Compared to national universities, branch campuses are perceived as more elite, progressive, and globally competitive (Lee 2021). Their physical presence not only confers a level of global prestige, but branch campuses also provide an opportunity for local elites to pursue outstanding higher education at home instead of going abroad. Therefore, branch campuses benefit host countries, such as Qatar and the UAE, by increasing the availability of quality higher education, expanding research activities, and providing opportunities for academic innovation and knowledge transfer (Agnew 2012; Wilkins and Rumbley 2018). Meanwhile, home institutions of branch campuses also benefit from increased overseas teaching and research opportunities, marketing/branding potential, financial gain, and greater influence in an emerging international higher education market (Wilkins and Huisman 2012; Zhan and Marginson 2023).

The varied motivations of the home institution and host country lead branch campuses to hold a dual identity (Lehman 2015). One identity rests in the home institution, while the other serves the interests of the host country (Lehman 2015; Zhan and Marginson 2023). Scholarly work has suggested that the institutional dual identity of branch campuses lies on a dichotomous spectrum of global integration and local responsiveness, related to dimensions of curriculum, staffing, and research (Shams and Huisman 2012; Knight 2015; Bellini et al. 2016). Shams and Huisman (2016) have further described the double forces and pressures that branch campuses face from home and the host, as the branch campuses have to simultaneously engage with and incorporate priorities, goals, identities, and context of the home institution and the host country.

The multiple goals and priorities pursued by both sides can, in practice, limit branch campuses' ability to have a social impact on their communities. Studies have found that branch campuses are likely to remain insular if neither the home campus nor the host country is motivated to consider aspects of social justice or identities and needs of different groups that comprise the expanded university community (Tikly and Barrett 2011; Lanford and Tierney 2016). On the other hand, if one of the motivations for building a branch campus involves a desire for social justice, then this is likely to be reflected in the institutional mission of the branch campus even if it may not be reflected in the institutional mission of the home campus (Lanford and Tierney 2016). While it is true that branch campuses may be scrutinized if they do not align with the institutional norms of the home campus, the mission statements of branch campuses can differ from those of the home campuses (Lanford and Tierney 2016). Indeed, it is not atypical for branch campuses to have distinctive mission statements that reflect a personalized curriculum, different departmental strengths, and distinct service to a particular local population (Wolfe and Strange 2003).

Therefore, when it comes to social justice-related issues such as sustainability and climate change action, it is likely that branch campuses face goals and priorities set by both sides, i.e., the home campus and the host country. For example, New York University is a signatory to the "Race to Zero" global campaign mentioned earlier. Both Georgetown University and New York University are institutional members of the Association for the Advancement of Sustainability in Higher Education (AASHE), and New York University has actually achieved a Gold rating in STARS (Sustainability Tracking, Assessment and Rating System). Additionally, both home institutions have established central offices of sustainability and appointed Chief Sustainability Officers, further signaling their organizational commitments to sustainability and climate change action. Simultaneously, both Qatar and the UAE have incorporated sustainability as part of their core pillars and priority areas of their national development visions (Bryde et al. 2015). Not only that, but both of these countries have also further demonstrated their national commitments to sustainability and climate change, with Qatar hosting the United Nations Climate Change Conference in 2012 (COP18) and the UAE hosting the same in 2023 (COP28).

The commitments of both the home campuses and host countries, then, presume that branch campuses in Qatar and the UAE would adopt and incorporate features of the home and the host into their own commitments and efforts toward sustainability. Given that the branch campuses in the Arab Gulf are often endorsed and funded by the state, rather than the home universities, we expect comparatively more references to the host country context than to their home campus. Given how embedded they are in the host country, the branch campuses may choose to frame their sustainability efforts as a part of and contribute to their respective host country's sustainability efforts. While acknowledging the home institutions' sustainability efforts, branch campuses may, in fact, leverage their global legitimacy and local prestige as a reputable institution to potentially lead the way in contributing to the host country's sustainability goals and efforts.

Regarding how branch campuses present and engage in their sustainability efforts, we draw upon neo-institutional theory (Meyer and Rowan 1977; DiMaggio and Powell 1983) and expect that sustainability narratives across branch campuses and their home institutions will be fairly similar and attuned to global rhetoric. Furthermore, we expect that the activities, initiatives, and programs that the branch campuses will engage in would be attuned to the kinds of activities and campaigns promoted by global efforts such as the "Race to Zero" campaign.

Just because the universities are attuned to global narratives and practices around sustainability, it does not mean that the practices on the ground will be identical. On the contrary, what happens on the ground likely differs based on the extent to which the university is embedded in the home institution and host country, in addition to how attuned the branch campus may be to the global, world culture around sustainability and climate change action. For example, if the branch campus is highly embedded within its home institution, then the branch campus likely mimics and implements various sustainability efforts of the home campus and adopts the home institution's sustainability goals and strategies. If, however, the branch campus is more embedded in the host country, then the branch campus may participate in events and activities that are promoted more widely in the host country and engage in partnerships with other local entities, emphasizing the importance of sustainability and climate change action for the local host country context. In what follows, we conduct an exploratory comparative study of two branch campuses in the Arab Gulf to examine how these dually embedded universities approach and implement sustainability initiatives in their respective contexts.

9.3 Research Context

We examine two American branch campuses in the Arab Gulf and their respective approaches to sustainability: Georgetown University Qatar (GUQ) and New York University Abu Dhabi (NYUAD). Using GUQ and NYUAD as our two case studies makes sense since they are both American branch campuses that are largely funded by and established in partnership with their respective host country governments. Simultaneously, both GUQ and NYUAD are affiliated with and maintain close relationships with their respective home campuses in Washington, DC, and New York City, so that students in the branch campuses also receive the "Georgetown" or "NYU" experience, should they choose to complete part of their studies at those sites. It is important to note that both branch campuses offer full undergraduate degrees, which are issued and recognized by the main campuses. However, NYUAD and GUQ differ in their organizational structures and relationships with the host campuses. NYUAD is a "replica campus," i.e., a full-scale research university and liberal arts and sciences college, offering both undergraduate and graduate programs, as well as research institutes (Miller-Idriss and Hanauer 2011). It is designed to fully replicate the educational experience of the home campus in Washington Square, including, to some extent, non-academic arenas such as student residential life and activities. On the other hand, GUQ is not a full-scale research university with both undergraduate and graduate programs. Rather, GUQ is a branch campus that offers solely degree programs within its School of Foreign Service and is part of a wider network of other American branch campuses that comprise Education City.

Since the structural setup of NYUAD and GUQ is similar yet different, we presume that these two campuses will vary in their degree of embeddedness within their respective home institutions and within their host countries. For instance, NYU publicly boasts about having a "campus, without walls," with NYUAD being one of its three degree-granting institutions. Therefore, it is possible that NYUAD is more closely embedded and connected with its home institution in New York City compared to GUQ. On the other hand, GUQ may be more closely connected to the local context of Education City in Qatar, especially in terms of sustainability efforts, as GUQ shares facilities with other branch campuses and offices located within Education City. Additionally, the fact that only select programs of Georgetown's School of Foreign Service are offered in Qatar may imply relatively lower levels of embeddedness on the main campus than would have been the case had it been a more comprehensive global campus, such as NYUAD.

In addition to the structure and subsequent relationship with the home institution, the GUQ and NYUAD are likely influenced in their sustainability orientations by developments in their respective host countries. Both Qatar and UAE have experienced global events that have led the two countries to be increasingly conscious of sustainability issues. In hosting the 2022 World Cup, Qatar has placed sustainability "at the heart of the FIFA World Cup Qatar 2022 from the very start," monitoring massive construction projects to minimize the impact on the environment (Supreme Committee for Delivery and Legacy n.d.). More pertinently, the UAE hosted COP28 and publicly declared 2023 as the Year of Sustainability to "inspire collective action through a nationwide commitment toward sustainable practices" (The United Arab Emirates Government Portal n.d.). These major world events, hosted by the host countries, provide a timely opportunity to examine how critical events and developments within the host country affect the sustainability efforts of branch campuses.

9.4 Data and Methods

Websites of organizations are both a form of representation and a type of narrative. They entertain, persuade, express a point of view, and provide a means to organize collective action and economic exchange (Powell et al. 2016). Such webpages serve expressive or symbolic functions, as well as instrumental ones. For all intents and purposes, they are the public face of an organization. Indeed, instead of visiting the organization in person or on the phone, customers and affiliates increasingly turn to the organization's website for information and to stay connected. Thus, it should come as no surprise that we relied heavily on university websites as a primary data source for our analyses.

The process of critically and systematically reviewing university websites to assess the presence or absence of specific features is a non-obtrusive data collection methodology commonly referred to in the literature as a "web census." This method has been employed successfully in other comparative higher education studies (see, for instance, Christensen and Gornitzka 2017; Gavrila et al. 2023). An additional advantage of using web pages as data sources is that they leave a permanent record online, which can be traced back over time. By utilizing the Wayback Machine functionality, which allows the public to access archived versions of websites, one can potentially determine when a page was first launched. In our case, we conducted a web census for both the home institutions and their branch campuses in the Gulf region, which are our two main case studies. Since both branch campuses have recently new sustainability webpages, we were able to pinpoint exactly when those webpages were first set up in order to figure how far back their public sustainability commitment goes.

We also supplemented the information collected directly from the university websites with information about the national and regional sustainability plans for Qatar and UAE, which we collected from other publicly available online sources such as Qatar Foundation, the State of Qatar's Government website, and the Official Portal of the UAE's Government. Last but not least, we acquired additional confirmatory information via email directly from university staff employed at those institutions. Taken together, our data sources for this chapter consist of website screengrabs, publicly available membership information, documents and reports, and private email exchanges. The data collected in this way was analyzed using qualitative methods, searching for recurrent themes (Cresswell and Cresswell 2018), as well as confirmatory and disconfirming evidence (Eisenhardt 1989) that would allow us to draw empirical conclusions about the propositions set forth in the conceptual framework.

9.5 Findings and Discussion

9.5.1 Symbolic Acknowledgement of Both Home Campus' and Host Country's Sustainability Efforts

Even though we found some baseline differences in how far back the parent websites for GUQ and NYUAD went (2018 versus 2014, respectively), the sustainability webpages at both branch campuses were equally new, with both having gone live as recently as 2022. Observation of the landing sustainability webpages of GUQ and NYUAD suggests that both universities are simultaneously attuned to their home campus' and their host country's sustainability efforts (refer to the excerpts in Table 9.1).

In comparing the ways that the two branch campuses emphasize and present their respective sustainability efforts; however, we find that the two campuses have different emphases, reflecting the ways that each of the campuses sees themselves as being connected to and embedded in the home campus and host country. Firstly, we observe that there are differences in the order in which the branch campuses situate their respective sustainability initiatives in context and the importance they place on different rationales for engaging with sustainability.

For instance, GUQ first situates its sustainability efforts relative to those of its main campus by citing their university president, before relating their efforts to the Qatari context. In addition to the website emphasis, we were able to ascertain the fact that GUQ maintains strong lines of communication with its home campus based on email exchanges with the sustainability staff at its home office in DC, which relayed the following when

Table 9.1 Excerpts from sustainability landing pages at the two selected branch campuses

GUQ	Sustainability efforts have been defined as practices that advance positive outcomes for people, the planet, and general prosperity and are aligned with Georgetown's mission of advancing the common good According to Sustainability at Georgetown Main Campus, Georgetown plays a leading role in addressing critical sustainability challenges through its academic mission, physical footprint, and financial operations. The university takes an integrated and holistic approach to sustainability, placing value on a "quadruple bottom line: people, planet, prosperity, and purpose" Here at Georgetown University Qatar, we want to lead with an example and as mentioned the University President, Dr John J. DeGioia, *"Environmental sustainability is one of the defining challenges of our time. At Georgetown we are committed to engaging in research, teaching and operational practices that address this challenge and help us minimize our impact on the environment."* Having said this, we face unique challenges in our Qatari (Regional) environment, but still we as Georgetown University Qatar can be committed to our environmental sustainability initiatives and plan As one of the four pillars of the Qatar National Vision 2030, Environmental Development is defined as *"Management of the environment such that there is harmony and balance between economic, social development and environmental protection"* We are determined to continue our work to raise awareness through sustainability initiatives within the GU-Q community and in partnership with external stakeholders and entities **Vision** We will aim to continue our work to instill in the Georgetown University Qatar community an appreciation and understanding of the impact of human choices on people around the world and the planet itself. Enabling our community to be force multipliers, taking their knowledge and passion to the home and beyond, promoting, influencing, and encouraging responsible, sustainable, and ethical practices in all aspects of their lives. "We need to care for our natural environment for it was entrusted to us by God to use with responsibility and respect for the benefit of humankind. If we nurture our environment, it will nurture us" Mozah bint Nasser Al-Misned **Mission** Georgetown University Qatar accepts that it must work to preserve the environmental sustainability of the planet, at all levels of its operations—in its own practice, as a participant in a community of practice, and as a participant in Qatar's National Vision 2030 Addressing climate change locally is not only an opportunity to reduce greenhouse gas emissions but also an opportunity to build a positive, community-based movement which results in increased empowerment, civic pride and improved quality of life *"We are building our Future Leaders"* (Georgetown University 2023)
NYUAD	NYU Abu Dhabi is committed to building and strengthening a university-wide culture of sustainability Through research and outreach, NYU Abu Dhabi is a thought leader in sustainability advocating the Sustainable Development Goals (SDGs) of the United Nations. Inspired by Sheikh Zayed's vision and legacy, we seek to have graduates across all divisions understand issues surrounding sustainability and how their work can contribute to a more sustainable world Under the leadership of the Office of Sustainability and Stewardship, students, faculty, and staff work closely to transform NYUAD into one of the region's leading campuses in sustainability. Through strategies that span every area of university life, we aim to reduce our greenhouse gas emissions, support teaching and research, and provide resources and guidance in establishing a culture of sustainability at NYU Abu Dhabi....The Office [of Sustainability and Stewardship] works with academic divisions to support sustainability teaching and learning; with the community to raise awareness around sustainability through events; with NYU's Office of Sustainability in New York to ensure adaptation of relevant common sustainability initiatives and strengthen relationships across all NYU campuses NYU Abu Dhabi is working to develop a comprehensive Climate Action Plan that will enable the university to measure its greenhouse gas emissions and set clear, measurable targets for reducing these as we move forward In alignment with the National Climate Change Plan of the UAE 2017–2050 and to support NYU's vision, NYUAD is bolstering action to reduce its carbon footprint and its impact on the environment NYUAD's Climate Action Plan will encompass all required reductions strategies to ensure that the targets set are achieved and that we are on track to succeed the university's aspirations (NYU Abu Dhabi, 2023)

asked about their working relationship: "We meet quarterly with the Georgetown Qatar campus to coordinate on sustainability. In between meetings, we also share resources, program ideas, data and other things across the two entities. It's a good working relationship." (Private email exchange, April 2023).

Based on their online presence, GUQ's vision with respect to sustainability appears to be both globally oriented and locally relevant. Their vision statement emphasizes instilling an appreciation and understanding of environmental impacts on people around the world and the earth, but does so by quoting Sheikha Moza, the Chairperson of Qatar Foundation and consort of Sheikh Hamad bin Khalifa Al Thani, the former ruler of Qatar.

On the other hand, NYUAD positions itself first and foremost in the global context, referring right off the bat to the U.N. Sustainable Development Goals and the founding father of the host country, Sheikh Zayed Al Nahyan. It is only after NYUAD describes its sustainability strategies for its own branch campus that it mentions how it will align with the home institution's goals to "ensure adaptation of relevant common sustainability initiatives and strengthen relationships across all NYU campuses." We were not able to receive additional confirmation via email about the communication channels between NYUAD's Office of Sustainability and Stewardship and the home campus' main Office of Sustainability, although we have reasons to believe they are just as connected as GUQ and its main office.

Despite the differences in the order of presentation (emphasis), one commonality is that both campuses view themselves as leaders who engage with sustainability through the totality of their research, teaching, and operations. Whether as a "thought leader" (NYUAD) or in "leading by example" (GUQ), the branch campuses express their desire to play an active and leading role in contributing to the building of a more sustainable world and planet. The global orientation and language suggest that the branch campuses, like their home institutions, are embedded in and influenced by global cultural norms that promote sustainability and sustainable development. Simultaneously, the place, i.e., the physical location, in which sustainability efforts take place, is the host country. Thus, it is unsurprising to find that the branch campuses are conscious of and incorporate national agendas around sustainability and set actions and plans to promote sustainability in ways that allow the branch campus to not only contribute to national sustainability initiatives but also lead and enable the campus community, including students and alumni, to promote and encourage sustainable practices.

9.5.2 Pragmatic Alignment of Select Sustainability Initiatives and Key Institutional Stakeholders

Next, we turn to the activities and initiatives undertaken by GUQ and NYUAD to examine how the branch campuses attempt to implement in practice the vision and mission publicly stated on their sustainability web pages. This will further allow us to understand how the particular sustainability activities and initiatives favored by each branch campus align with global practices and are responsive to their own local, national, or regional context.

Table 9.2 provides examples of activities and initiatives engaged by GUQ and NYUAD to promote and encourage sustainability within and beyond their branch campuses. To frame these activities and initiatives, it is worth noting from the outset that the Office of Facilities Management oversees sustainability efforts at GUQ, while the Office of Sustainability and Stewardship, mentioned previously, oversees sustainability efforts at NYUAD.

The overview of sustainability initiatives captured in Table 9.2 shows that both GUQ and NYUAD are attuned to global practices surrounding the observance of symbolic sustainability celebrations (e.g., Sustainability Week, Earth Day, etc.). Both branch campuses also engage in campus-focused activities that have become "standard practice" among sustainability-conscious universities everywhere, such as recycling, waste minimization, and water conservation

Table 9.2 Examples of key activities and initiatives engaged by GUQ and NYUAD

	Campus community and operations	Research and teaching	External engagement
GUQ	• Observed green earth day and world facilities management day • Recycling and food waste awareness campaigns • WATER and energy conservation • Sustainability dashboard • Tree planting initiative • Student interest groups (e.g., Sustainability Club, Islamic Environmentalism)	• "Introduction to Environmental Humanities" course offering and associated "Doing Environmental Humanities in Doha" public course blog • Sustainability-related scholarship published by GUQ faculty • Panels and lectures promoting sustainability literacy and awareness of local environmental issues	• Participation in Qatar's Sustainability Week (2022) • Participated in sustainability campaigns through the Qatar Green Business Council (QGBC) (2019) • Shared sustainability practices with Other institutions inside and out Education City to enable them to "Go Green"
NYUAD	• Observed sustainability week • Recycling and waste minimization programs • Sustainable dining • Green buildings and green transportation • Tree planting • Student interest groups and committees (e.g., (*Ecoherence*, Green House, Sustainability Committee) • Green workplace Certification program • Green grants program • Climate Action Plan (CAP) (currently under development)	• Green Capstone • Public talks on sustainability-related issues and on research conducted by the NYUAD Institute • Various labs and research centers on sustainability-related topics (e.g., Arabian Center for Climate and Environmental Sciences, Water Research Center)	• Participation in the "Road to COP 28" activities hosted at Expo City (2023) • NYUAD-led Industry-Academia Sustainability Dialogues (2023) • Chair of the Universities Climate Network (UCN) in the UAE (a network comprised of 12 UAE-based higher education institutions to encourage youth engagement in the lead up to COP 28)

programs. Another common practice promoted by both GUQ and NYUAD was tree planting; however, the two branch campuses differed slightly in the way they reported their efforts in that regard: GUQ mentioned they have "planted trees for each graduation class" as part of their sustainability achievements, while NYUAD reported that they initiated "UAE's largest plantation of mangrove trees as part of the University's commitment to sustainability and environmental conservation." (NYUAD 2022). While the campus community was referenced in both instances, NYUAD has a decidedly more outward-facing angle in the way they conduct some of their sustainability initiatives on the ground.

Moreover, as can be expected of any US-affiliated campuses, both GUQ and NYUAD support student-led initiatives to raise awareness of daily environmental impact among their student bodies and allow students to engage with issues of sustainability on their own terms, at times in conversation with other salient social issues and identities (see for instance, the Islamic Environmentalism club at GUQ or the *Ecoherence* student club and the Green House at NYUAD). Additionally, NYUAD manages a Green Grants Program (in collaboration with their New York campus), which is awarded to "projects seeking to advance campus sustainability initiatives or enact social-entrepreneurial ventures." These grants are open to current NYU students, faculty, and staff at any of NYU campuses or global sites, while "alumni, community organizations, and interested neighbors are permitted and encouraged to collaborate as members of project teams." Once again, this additional sustainability program aimed primarily at their internal constituents also suggests that NYUAD tries to expand its sustainability reach beyond the campus walls as much as possible.

Another commonality between how GUQ and NYUAD implement their public sustainability commitments in practice is the tracking of concrete sustainability data and metrics related to the environmental footprint of their campuses. GUQ uses an in-house sustainability dashboard ("pending peer review") to collect and report relevant data and trends over time. Meanwhile, NYU has developed an elaborate methodology for calculating its greenhouse gas emissions inventory across all campuses, which the NYUAD campus plans to incorporate into its future Climate Action Plan (CAP), currently under development. Although this is considered a strategic document primarily used for internal decision-making, NYUAD references the local context when justifying this particular initiative. In particular, they strive to align their future CAP with the UAE's National Climate Change Plan (2017–2050).

When it comes to folding sustainability into their broader teaching and research mandate, GUQ and NYUAD follow similar strategies: They both take an active role in promoting sustainability literacy and innovation on campus and within their local communities by organizing public talks and panels to report research and scholarship on sustainability-related topics conducted by their faculty and students. At the curricular level, GUQ has been offering a high-visibility course on "Introduction to Environmental Humanities," which resulted in an expansive public project titled, *Doing Environmental Humanities in Doha: A Georgetown-Qatar Exploration.* On their part, NYUAD offers a wide selection of sustainability-related courses and ample opportunities to connect with faculty and labs that conduct research in this area. Additionally, every student at NYUAD is expected to conduct a Capstone project in their fourth year, a demanding, year-long endeavor that aims to produce a significant piece of research or creative work. Recently, NYUAD has introduced a Green Capstone option for its students through the Tomoh Fellowship Program, which it claims will be a "key contributor to NYUAD's mission of developing a sustainable, knowledge-based economy in Abu Dhabi."

When it comes to the external engagement of branch campuses around sustainability, we find that both GUQ and NYUAD participate in sustainability initiatives of their respective host countries. The branch campuses are not just passive participants but active participants, hosting events and activities on campus and for community members as part of national initiatives such as Qatar's Sustainability Week and UAE's Road to COP 28 events. The branch campuses' involvement in local and national sustainability initiatives is consistent with their respective visions to contribute to the sustainability efforts of their host countries. Not only so, but it is also reflective of how a part of their identity serves the interests of the host country, as the branch campuses face pressure from the host countries to engage with and incorporate the goals and priorities of the host country (Lehman 2015).

In fact, we find that NYUAD's outward-facing sustainability initiatives are aligned with and launched in coordination with major events and activities of its host country, i.e., the UAE. NYUAD established the Office of Sustainability and Stewardship in 2022, in anticipation of the UAE hosting COP28 in 2023 and has framed and legitimated its increased emphasis on sustainability initiatives on campus around UAE's sustainability efforts. For example, in launching NYUAD's first Net Zero Now Week, Mariet Westermann, the Vice Chancellor of NYUAD, stated,

> We are redoubling our efforts in this UAE Year of Sustainability. NYUAD will play an important role when the country hosts COP28 in November and December (Office of the Vice Chancellor 2023).

Such timeliness of NYUAD's sustainability efforts suggests that the branch campus's sustainability efforts are closely aligned with and influenced by its external, local environment. A major world event such as COP28, being held in the UAE, combined with UAE Year of Sustainability, provides an impetus for the branch campus to further its sustainability efforts and position itself as a local, university leader for sustainability as a "socially embedded" and "socially responsible" university (Ramirez 2020). Sustainability serves as a motivating force for the branch campus to remain engaged with the host country, rather than remaining insular, as it is a social justice issue of great interest to both the home campus and the host country. Hence, by strategically positioning itself as a "global" leader in sustainability as a global (branch) campus in its host country, it can manage its public image as an institution that is locally embedded. This, then, allows the branch campus, such as NYUAD, to address critiques of being a "bubble" that is not well integrated into the local UAE context (Cai and Hall 2015; Lanford and Tierney 2016) while furthering the home institution's sustainability goals and image as a "green campus."

9.6 Limitations and Future Research

As is commonly the case for organizations engaged in the sustainability movement, there is always a risk that university campuses may overstated their public commitment to sustainability in a manner reminiscent of "greenwashing," a term more commonly used in the private sector. Indeed, it is not only GUQ and NYUAD that are theoretically at risk of engaging in greenwashing practices but their home institutions and host countries as well. While it is beyond the purview of this chapter to assess the extent to which any of those entities may have engaged in greenwashing, intentionally or accidentally, we do want to flag this phenomenon as a potential limitation of our chapter, since websites are the prime outlet through which greenwashing may occur and we have relied heavily on information collected from the two universities' websites. To mitigate this risk, we divided our analysis between declarative statements regarding sustainability and an assessment of the specific sustainability initiatives and programs that the two branch campuses purportedly engage in. However, we cannot confirm the extent to which these initiatives were implemented in practice and the impact they have had on the ground. Since ours is exploratory, in-depth interviews and/or participant observation could be used in future to ascertain these preliminary findings.

Our research context has also been limited by our choice of case studies. In order to let some selection factors vary while keeping others constant across the two case studies, we have chosen to focus on two American branches in two different Gulf countries. Given that there are many different branch campuses in the region, one could envision re-doing or expanding this study by examining other foreign or Western (non-American) branch campuses. Alternatively, comparing the sustainability efforts of branch campuses to those of national universities would provide greater insights into how different levels of local (and global) embeddedness influence the ways universities engage in and promote sustainability.

9.7 Conclusion

In this chapter, we observed how two American branch campuses in the Arab Gulf present and engage in their sustainability efforts. We find that sustainability narratives across the branch campuses and their home institutions are generally similar and aligned with global rhetoric. We also find that the activities, initiatives, and programs that the branch campuses are attuned to the kinds of activities and campaigns promoted by global efforts such as the "Race to Zero" campaign. Simultaneously, we find that the branch campuses are attuned to sustainability initiatives and goals of their host countries and envision themselves as leaders and proactive participants in contributing to the sustainability efforts of their host countries. We further find that the branch campuses leverage locally held world events, such as the 2022 FIFA World Cup and COP28, to position themselves as local university leaders for sustainability. As is the case for NYUAD, the UAE hosting COP28 and declaring 2023 as the Year of Sustainability became an impetus and motivator to accelerate its sustainability efforts in ways that are more embedded and interactive with the local host country context. There are, of course, concerns for "greenwashing." Nonetheless, it appears that goals such as sustainability, which is one that is globally encouraged and of interest to both the home institution and the host country, provide the branch campuses an opportunity to leverage their dual embeddedness to position themselves as a local (and global) leader.

References

Agnew M (2012) A false dichotomy of serving either the local or the global community and its impact on internationalisation of the university. J High Educ Policy Manag 34(5):473–489. https://doi.org/10.1080/1360080X.2012.715997

Altbach PG, de Wit H (2020) Rethinking the relevance of international branch campus. Int High Educ 101:14–16

Becker R (2015) International branch campuses: new trends and directions. Int High Educ 58:3–5

Bellini N, Pasquinelli C, Rovai S, Tani S (2016) The local embeddedness of foreign campuses: The case of Tongji University in Florence. J Stud Int Educ 20(4):371–385. https://doi.org/10.1177/1028315316656941

Bryde D, Mouzughi Y, Al Rasheed TF (2015) Sustainable development challenges in the Arab states of the Gulf. Gerlach Press, Berlin, Germany

Cai L, Hall C (2015) Motivations, expectations, and experiences of expatriate academic staff on international branch campuses in China. J Stud Int Educ 20(3):207–222. https://doi.org/10.1177/1028315315623055

Christensen T, Gornitzka A (2017) Reputation management in complex environments—a comparative study of university organizations. High Educ 30:123–140. https://doi.org/10.1057/s41307-016-0010-z

Creswell JW, Creswell JD (2018) Research Design: Qualitative, Quantitative, and Mixed Methods Approaches. Sage, Los Angeles.

DiMaggio PJ, Powell WW (1983) The iron cage revisited: institutional isomorphism and collective rationality in organizational fields. Am Sociol Rev 48(2):147–160. https://doi.org/10.2307/2095101

Disterheft A, Caeiro S, Azeiteiro UM, Leal Filho W (2013) Sustainability science and education for sustainable development in universities: a way for transition. In: Caeiro S, Filho W, Jabbour C, Azeiteiro U (eds) Sustainability assessment tools in higher education institutions. Springer, Cham

Eisenhardt KM (1989) Building theories from case study research. Acad Manag Rev 14(4):532–550

Figueiredo PN (2011) The role of dual embeddedness in the innovation performance of MNE subsidiaries: evidence from Brazil. J Manage Stud 48(2):417–440. https://doi.org/10.1111/j.1467-6486.2010.00965.x

Filho WL (Ed.) (2000) Sustainability and student life (2nd ed.). Peter Lang Publishing, Lusanne, Switzerland

Frank DJ, Hironaka A, Schofer E (2000) The nation-state and the natural environment over the twentieth century. Am Sociol Rev 65(1):96–116. https://doi.org/10.2307/2657291

Gavrila SG (2023) Green is the new ivory: the rise of the sustainability imperative in US higher education. [Doctoral Dissertation, Stanford University].

Ibnouf A, Dou L, Knight J (2014) The evolution of Qatar as an education hub: moving to a knowledge-based economy. In: Knight J (ed) International Education Hubs. Springer, Dordrecht, pp 43–61

Jones DR (2012) Looking through the "greenwashing glass cage" of the green league table towards the sustainability challenge for UK universities. J Organ Chang Manag 25(4):630–647. https://doi.org/10.1108/09534811211239263

Knight J (2015) International universities: misunderstandings and emerging models? J Stud Int Educ 19(2):107–121. https://doi.org/10.1177/1028315315572899

Koch N (2018) Green laboratories: university campuses as sustainability 'exemplars' in the Arabian Peninsula. Soc Nat Resour 31(5):525–540. https://doi.org/10.1080/08941920.2017.1383546

Lane JE, Pohl H (2017) Is there a benefit to importing branch campuses? research capacity in Abu Dhabi. Int High Educ 89:14–16

Lanford M, Tierney WG (2016) The international branch campus: cloistered community or agent of social change? In: Collins C, Lee M, Hawkins J, Neubauer D (eds) The Palgrave handbook of Asia pacific higher education. Palgrave Macmillan, New York

Lee SS (2021) A precarious balancing act: globalization, political legitimacy, and higher education expansion in Qatar and the UAE. Contemp Arab Affairs 14(1):113–133. https://doi.org/10.1525/caa.2021.14.1.113

Lehman J (2015) Opportunities and challenges of a "Double Identity" model. China Annual Conference for International Education, Beijing, China

Meyer JW, Rowan B (1977) Institutionalized organizations formal structure as myth and ceremony. Am J Sociol 83(2):340–363. https://doi.org/10.1086/226550

Miller-Idriss C, Hanauer E (2011) Transnational higher education: offshore campuses in the Middle East. Comp Educ 47(2):181–207. https://doi.org/10.1080/03050068.2011.553935

Powell WW, Horvath A, Brandtner C (2016) Click and mortar: Organizations on the web Research in Organizational Behavior 36101–120. https://doi.org/10.1016/j.riob.2016.07.001
Ramirez FO (2020) The socially embedded American university: intensification and globalization. In: Engwall L (ed) Missions of universities over time. Springer, New York, pp 131–162
Sahlin K, Wedlin L (2008) Circulating ideas: translation, imitation and editing. In: Greenwood R (ed) Handbook of organizational institutionalism. Sage, pp 218–242
Shams F, Huisman J (2012) Managing offshore branch campuses: an analytical framework for institutional strategies. J Stud Int Educ 16(2):106–127. https://doi.org/10.1177/1028315311413470
Shams F, Huisman J (2016) The role of institutional dual embeddedness in the strategic local adaptation of international branch campuses: evidence from Malaysia and Singapore. Stud High Educ 41(6):955–970. https://doi.org/10.1080/03075079.2014.966669
Sylvestre P, McNeil R, Wright T (2013) From Talloires to Turin: a critical discourse analysis of declarations for sustainability in higher education. Sustainability 5(4):1356–1371. https://doi.org/10.3390/su5041356
Tikly L, Barrett AM (2011). Social justice, capabilities, and the quality of education in low income countries. Int J Educ Develop 31(1):3–14. https://doi.org/10.1016/j.ijedudev.2010.06.001
Vora N (2018) Teach for Arabia: American universities, liberalism, and transnational Qatar. Stanford University Press, California, USA.
White S (2015) Gulf state branch campuses: global student recruitment. Int High Educ 58:4–6
Wilkins S, Huisman J (2012) The international branch campus as transnational strategy in higher education. High Educ 64(5):627–745. https://doi.org/10.1007/s10734-012-9516-5
Wilkins S, Rumbley L (2018) What a branch campus is: a revised definition. Int High Educ 2(93):12–14. https://doi.org/10.6017/ihe.0.93.10416
Wolfe JR, Strange CC (2003) Academic life at the franchise: Faculty culture in a rural two-year branch campus. Rev High Educ 26(3):343–362. https://doi.org/10.1353/rhe.2003.0006
World Commission on Environment and Development [WCED] (1987) Our common future
Zhan T, Marginson S (2023) Institutional dual identity in research capacity building in IBCs: the case of NYU Shanghai. High Educ 87(2):471–490. https://doi.org/10.1007/s10734-023-01017-9

Seungah Sarah Lee is an Assistant Professor at HEC Paris. Her main research interests lie in the areas of organizational studies, entrepreneurship, higher education, and models of national development. She holds a PhD from Stanford University.

Gabriela Reynell (Gavrila) is a Program Specialist at the Berkeley Global Access Program at UC Berkeley. Her major research interests lie in the area of higher education, sustainability, community partnerships, diversity, equity, and inclusion. She holds a PhD from Stanford University.

10 The Role of Research Centers in Supporting Higher Education and Sustainable Development: Evidence from Sultan Qaboos University

Omer Hamad

Abstract

Climate change (CC) is an international phenomenon and serious challenge to the humanity globally. The rise in global temperature, caused by greenhouse gas emissions, poses a significant threat to human life. CC, with its impacts, is becoming increasingly evident and predictable. Floods, cyclones, storms, earthquake, droughts, and desertification are all considered "climate-related disasters" that damage infrastructure, lead to loss of life, and hinder socio-economic development. To ensure sustainable progress, it is crucial to study, identify, and effectively manage the impacts of climate change. Higher education institutions (HEIs) play a vital role in addressing CC and advancing sustainable development. Their contributions to CC adaptation and mitigation efforts are essentials. In this context, universities in Gulf Cooperation Council (GCC) countries must enhance their engagement by educating students and conducting research on global warming and CC. This paper examines Climate Change Research (CCR) projects undertaken by research centers at Sultan Qaboos University (SQU), highlighting their efforts to CC mitigations and adaptation strategies.

O. Hamad (✉)
Sultan Qaboos University, Muscat, Oman
e-mail: omerhi@squ.edu.om

Keywords

Climate change · Oman · Research · Research centers · Sultan Qaboos University

10.1 Introduction

Climate change has become one of the most worrying issues internationally due to the disasters, scale of destruction, and panic it leaves behind. In 2022, a dry wave hit Europe, leading to a significant decrease of rivers water levels and drying up of some lakes (Symanski 2022; Cammalleri et al. 2020; Toreti et al. 2022).

Scientists warned that global warming is advancing seriously beyond normal conditions and beyond the ability of the ecosystem to deal with it. Hence, an international conference was held in 1985 to assess the role of carbon dioxide and other gases in global warming and climate changes. Scientists, therefore, called for mechanisms to curb climate change (Yamin and Depledge 2004; IPCC 2007). These global efforts culminated in the Vienna Convention for the Protection of the Ozone Layer in 1985, then the Montreal Protocol to protect the ozone layer in 1987 and the Kyoto Convention in 1997. All of these conventions and protocols aimed at finding common solutions and an international agreement in which all countries cooperate to limit climate

K. Al'Abri and E. Buckner (eds.), *Higher Education and Sustainable Development in the Arab Gulf States*, Sustainable Development Goals Series, https://doi.org/10.1007/978-3-032-19559-3_10

change and its impact on the globe (Battaglini and Harstad 2016; Harstad 2016).

Nevertheless, despite of these ongoing international efforts, climate change continues to pose a major challenge, largely due to lack of agreement and differences among countries in their perspectives on the issue and the level of commitment they demonstrate (Kates 2018). Reports and scientific research also showed a rise in numbers, strength, and intensity of hurricanes in the Atlantic and Indian Oceans, and the scale of destruction left by them has increased as well. In 2019 there were approximately 55 hurricanes, in 2020 there were 46 hurricanes, in 2021 there were approximately 37 hurricanes, and in 2022 there were 40 hurricanes around the world (Salas 2024). These hurricanes, as statistics reveal, have also had negative social impacts, as they led to the displacement of 2,934,000 people in 2017 and 24.2 million people in 2016 (IDMC 2017). In 2020, natural disasters led to 12,886 deaths around the world (Hannah et al. 2022).

While these statistics reflect the global impacts of natural disasters, their repercussions are equally evident at the regional and local levels. The Sultanate of Oman has been no exception from such challenges. Oman is located in the southeastern part of the Arabian Peninsula, bordered by the United Arab Emirates to the northwest, Saudi Arabia to the west, and Yemen to the southwest. It also has a coastline along the Arabian Sea, Gulf of Oman, and the Strait of Hormuz. Oman is known for its diverse landscapes, from the desert of the Rub' al Khali to the lush green hills of Dhofar. The total area of Oman is 309.5 thousand square kilometers and, according to the Annual Health Report (2021), its total population is 4,527,446 people.

Oman is one of the countries in the Arabian Peninsula that is particularly vulnerable to the impacts of climate change due to its location in a hot and arid region. As Al-Saidi and Al-Agib (2018) mentioned, climate change is a significant environmental challenge that is affecting Oman in various ways, including rising temperatures, changes in precipitation patterns, sea level rise, and impacts on infrastructure, economy, and society. In addition, Charabi (2013) stated that the threat from the potential impacts of climate change is growing with the recent tropical cyclones that had affected the country and caused loss of life and substantial damage throughout the coastal areas of Oman. This highlights the significance of studying the effect of these climate changes on the Sultanate of Oman, which has been hit by a number of hurricanes, including Hurricane Gonu in 2007, Hurricane Phet in 2010, Hurricane Mekunu in 2018, and Hurricane Shaheen in 2021, causing human and material damage.

Higher education institutions have a fundamental role to play in advancing sustainability. As Jones (2012) and Fazey et al. (2021) mentioned, universities have the potential to stimulate and accelerate societal transformation toward sustainability and in achieving sustainable development goals, as well as conducting scientific research on environmental challenges.

Universities have the potential and opportunities to benefit from large intellectual, financial, material, and labor resources to drive transformation and reduce the risks of climate change and devastating human suffering (Gamoran 2018; Ahmed et al. 2022). They are the right place to understand the nature of change and its dimensions and effects and work to reduce it and build a sustainable future (Schoch-Spana et al. 2019). They can also manage multidisciplinary research teams and can partner with the professional community from inside and outside the university.

Climate change issues need insights and in-depth studies by scientists and experts in environment, economics, politics, law, science, education, sociology, medicine, agriculture, engineering, and biology (Garmody 2009; Bai et al. 2016). Hence, this paper investigates the efforts made by research centers at Sultan Qaboos University to address climate change issues in the Sultanate of Oman.

This chapter examines Climate Change Research (CCR) projects undertaken by research centers at Sultan Qaboos University (SQU), which has 14 research centers.

10.2 Defining and Classifying Research Gaps

A research gap can be identified as any issue that has not yet been adequately addressed or investigated by researchers. Climate change, encompassing environmental, economic, social, and political dimensions, remains under researched, particularly within regional and local contexts. Identifying research gap is therefore essential for advancing knowledge and highlights areas that require further investigations and address the unique challenges posed by climate changes.

A research gap, as defined by Baako et al. (2022) is the piece of information or knowledge in the research literature regarding an area of research interest that has not yet been explored or is under-explores. It is referred to as "unexplored or underexplored areas that have scope for further research" (Qureshi 2019). It also means unanswered questions or unsolved problems.

Al-Saraf (2022) stated that "a research gap can also exist when there's already a fair deal of existing research, but where the findings of the studies pull in different directions, making it difficult to draw firm conclusions." Müller-Bloch and Kranz (2014) integrated different definitions and stated that "a research gap arises when there is a gap in sets of information that is derived from a literature synthesis and requires further research to be resolved." Thus, research gap in its broad meaning includes issues lacking and needing scientific research and investigation. For example: questions to be answered, problems to be solved, ideas to be explained, old research to be updated, and a country or a population to be studied.

There is no agreed upon set of types of research gaps, as different researchers identified different types. For example, Miles (2017) named seven research gaps, while Robinson et al. (2011) identified five. Miles (2017) proposed and defined the following seven types of research gaps:

(1) Evidence Gap: Arises if new research finding contradicts widely accepted conclusions.
(2) Knowledge Gap: Occurs when there are two settings: First, knowledge may not exist in the actual field to theories and literature from related research domains. Second, it might be the case that results of a study differ from what was expected.
(3) Practical-Knowledge Conflict Gap: Arises when the actual behavior of professionals is different from their advocated behavior.
(4) Methodological Gap: Deals with the conflict that occurs due to the influence of methodology on research results.
(5) Empirical Gap: Addresses conflicts that no study to date has directly attempted to evaluate a subject or topic from an empirical approach.
(6) Theoretical Gap: Deals with the gaps in theory of prior research.
(7) Population Gap: In this gap the population is not adequately represented or under-researched in the evidence base or prior research (e.g., gender, race/ethnicity).

The five types of research gaps delineated be Robinson et al. (2011) are: population, intervention, comparison, outcomes, and setting. Likewise, Al-Saraf (2022) highlighted many different types of research gaps, the most common of which are:

(1) The **Classic** Literature Gap: which means there's a **new concept or phenomenon** that hasn't been studied much, or at all.
(2) The **Disagreement** Gap: which emerges when there are **contrasting or contradictory findings** in the existing research regarding a specific research question (or set of questions).
(3) The **Contextual** Gap: which exists when there's already a decent body of existing research on a particular topic, but an **absence of research in specific contexts**, such as population, country, and specific time frame.
(4) The **Methodological** Gap: which emerges as a result of the **research methodology or design** of existing studies.

According to Alex (2022), the identification of research gaps is a crucial step in scientific research. It generates subsequent research that is informed by input from previous research studies, thereby contributing to the growth of knowledge. Moreover, understanding areas of uncertainty within the research problem, which is vital for refining research focus and addressing unsolved challenges. It also facilitates the establishment of clearly defined research problem, ensuring that that the research topics are well-grounded and purposeful. Finally, identification of research gaps help determine the scope of funding opportunities, as research that addresses underexplored areas can attract financial support.

10.3 Methodology

The research methodology used for this paper includes a review of publications, reports, and research projects conducted at SQU research centers, as well as primary qualitative data collection through in-depth interviews with the Research Centers' directors and researchers. Expected findings include the identification of relevant and significant research conducted by the research centers in SQU to support climate change adaptation and mitigation efforts. Research gaps in climate change and related issues will be identified, and suggestions for the development of CCR policy in SQU will be provided as well.

Sultan Qaboos University recognizes research as one of its main functions and therefore, it promotes and facilitates research. This has been reflected in the establishment of 14 research centers. A brief summary of these research centers is provided below.

(1) ***Earthquake Monitoring Research Center***

Established in 2001, this center offers services to the community through research studies related to earthquake monitoring. It focuses on measuring the potential threat of earthquakes in ways that can assist in discovering the best means of limiting their effect and potential damage. The center operates through a network of twenty stations distributed in various areas in the sultanate. It also monitors seismic waves coming from the seismic zones.

(2) ***Environmental Studies Research Center***

Established in 2000, this center plays an important role in organizing and coordinating studies and research related to environment and in raising awareness of environmental issues and topics related to the sultanates. It collaborates with various government agencies with initiatives that contribute to the conservation of natural resources that can help achieve sustainable development. The center contributes to the formation of environmental policies and offers assistance to organizations and stakeholders that seek its input.

(3) ***Oil and Gas Research Center***

Founded in 2002, the center encourages and supports scientific research and strategic studies in areas related to oil and gas and contributes to transferring, localizing, and developing unconventional and innovative technologies. It is also concerned with training Omani nationals in research related to oil and gas.

(4) ***Earth Sciences Research Center***

This center aims to investigate ideas and develop pioneering strategies for environmental research in various areas of earth sciences, especially in oil, gas, and mining. The center manages a database of Oman's geological features, serving as the primary source of geological data and related investment opportunities in. In addition, it provides expert advice on the exploration and exploitation of mineral resources.

(5) ***Sustainable Energy Research Center***

Established in 2017, this center covers research topics that include energy policy and strategy issues, renewable energy, integrated power systems, energy management and rational use, and their economic and social aspects. It aims to unify individual energy researches efforts at SQU under one umbrella that encompasses all areas of sustainable energy. It also aims to provide energy consultancy services across all sectors and to assist and support master's and doctoral students in conducting their research.

(6) ***Nanotechnology Research Center***

Established in 2017, this center solidifies the presence of SQU regionally and internationally in the fields of nanotechnology and serves as the primary house of expertise in nanotechnology research in the Sultanate. It provides research leadership and builds capacity in the area of nanotechnology and contributes to sustainable development at national, regional and global levels.

(7) ***Remote Sensing and GIS Research Center***

This center was established in 2002 to support educational and research activities related to remote sensing, aerial photography and GIS within the university. The center coordinates between the university, ministries, industry and other institutions inside and outside the Sultanate. It also serves as a national resource in this field, providing strategic services, consulting, training, and promoting international cooperation.

(8) ***Water Research Center***

Established in 2002, this center studies mechanisms to develop the water sector in Oman and coordinates research aimed at developing water resources and protecting them from pollution and depletion. The center explores research priorities in areas of water, while also offering information to researchers and providing research services such as consultancies and analysis through available university resources.

(9) ***Excellence in Marine Biotechnology Research Center***

Established in 2017, this center act as a focal point for university-wide research in marine biotechnology. It promotes capacity building and the optimal use of Omani marine resources through high quality research, technical support, training, conferences, and consultations. The center seeks sustainable development that will lead to future growth and prosperity in this area in Oman, which is important due to the Oman's long coastline and diverse marine resources that are currently unexplored and exploitable.

(10) ***Medical Research Center***

This center was established in 2018 to produce high quality medical research of a broad and specialized nature that is conducted in well-equipped laboratories. It also focuses on prompting research collaboration between stakeholders within Oman and across the country and the region.

(11) ***Communication and Information Research Center***

It was established in 2001 to conduct joint research related to information and communication technology. The center works in partnership with and liaison between government institutions, SQU, and industry. In addition, the center implements research strategies, which meet the information and communication technology requirements of Oman in both the short and long terms.

(12) ***Humanities Research Center***

Established in 2010, this center seeks to promote and play a leading role in the field of humanities through the advancement of research work and the development of tools to face the rapid global challenges that affect all aspects of life.

(13) ***Omani Studies Research Center***

Established in 2009, this center focuses on conducting research studies and surveys related to political, economic, social, and cultural development in the Sultanate of Oman. In order to play a role in achieving Oman Vision 2040, the center encourages research competencies and seeks to build a generation of researches while spreading a culture of research and scientific publications.

(14) ***Innovation and Technology Transfer Center***

It was established in 2018 as mandated by SQUs strategic Plan 2016–2040 which underline the importance of establishing an independent center for innovation. The center contributes to Oman's social and economic development by supporting young Omanis in learning about, and engaging in, every aspect of innovation and entrepreneurship. It also aims to help fulfill national objectives related to strategic socio-economic development and contributes to the diversification and sustainability of Oman's economy.

10.4 Climate Change Research at SQU Research Centers

Unstructured interviews were conducted with Directors of four research centers to explore the extent to which these centers are engaged in conducting research and studies on climate change issues. Collected data were analyzed and reported as follows.

R1: The Director of the Environmental Research Centre (ERC), who has led the center for more than 6 years, stated that: We defined research clusters. We had almost 6 research clusters, one of them focused on climate change (CC). Under this topic (CC), the Centre attracted 3 research consultancies from international organizations. He named three research grants from Green Climate Fund (GCF). "One was about institutional readiness for dealing with climate change. The second research grant was on national adaptation plan to help Sultanate of Oman to adopt to climate change. The third research grant aimed to develop strategy for low Carbone transportation in Oman."

Speaking on CC research at SQU, he explained that: "when we talk about Climate Change at Sultan Qaboos University as a research track, we don't have this recognition. At SQU we don't have research priority focusing on CC, and this is one of the weaknesses. Now they have a new policy, that direct research to provide services to different government entities. This means these government entities suggest what their problems and challenge are, and the competition for university research funds is based on these topics. As a result, the priority for funding is based on these government entities. And we don't find any priority related to CC."

However, R1 added, due to the university ranking requirements, the University must submit a report every year showing the achievement of the university in regards to climate change. This is required by QS ranking. They ask for two aspects in terms of teaching and learning. If the university has any courses on climate change (which we don't have). The only department that offers this course is our department. One about climatology and another elective on climate change.

R2: The Director of Earthquake Centre stated that they cannot say there is a lack of attention to climate change, based on the fact that earthquakes contribute very little to climate change. Earthquakes occur at deep depths, at least five kilometers below the Earth's surface, and can reach up to six hundred kilometers, whereas climate change occurs between ten to fifteen kilometers. This means that earthquakes have a minimal impact on climate change or even contribute to it. Secondary hazards of earthquakes can amplify or increase the effects of climate change.

The climate change resulting from earthquakes is manifested in temperature rise, leading to melting of ice, rising sea levels, and the occurrence of tsunamis. The damages to infrastructure, particularly oil facilities, result in emissions that can have short-term or long-term changes. As there is no direct link, there is no research gap because the seismic monitoring center primarily focuses on earthquakes, and we have not considered climate change before.

R3: The Remote Sensing Centre Director said that there is no research gap in the center when it comes to studying climate change. This is evident in the center's utilization of the latest technologies in analysis, evaluation, and environmental monitoring in the fields of oil, minerals, pollution identification, weather conditions, as well as the opening of the Seventh Center of Excellence for the World Meteorological Organization in 2006, which analyzes aerial images and predicts any changes.

R4: The Oman Studies Center Director acknowledged that according to the objectives of the Center, there is no research gap. The center has conducted several studies on environmental issues, such as sustainability, climate change, and clean energy. Some of these studies have even had global recognition in the field of educational energy studies. They have been published in Arabic and English journals. Additionally, many postgraduate students at the master's and doctoral levels have focused their research on the environment, environmental issues, and environmental sustainability.

The interviews also addressed the reasons for the limited interest of the university's research in conducting in research dealing with climate change.

According to Dr. Charabi (former Director of Environmental Studies Research Center), it was clear that climate change research wasn't a concern for researchers in Oman. He stated that: "After Gonu Cyclone, the government started developing interest in CC. And one of the first responses to CC was the development of the Ministry of Environment and Climate Affairs. I explained to them at that time that CC is not a question of believe, it is rather a question of scientific fact. ... The CC is happening, it is not in the future, it is happening now, and there is an acceleration in the level of the change for different areas impacted by CC. However, here in Oman the priority is sustainable development, enhance or recovery, energy sector. They are not focusing on climate. Why? because they consider here is already a hot climate and if temperature increases by 2 or 3 degrees, it will not have any impact. But this not the case. There is an impact on fisheries, water resources, agriculture crops, health (yellow fever, malaria) specially ...."

R2 (Earthquake C): Climate change does not fall within the expertise of seismic monitoring, which has led to a lack of interest and focus on climate change-related research and studies in earthquake operations. This lack of interest stems from the center's specialization not being directly related to climate change. It has specific obligations and focuses on its assigned areas of expertise. As mentioned before, there are other centers that are dedicated to studying climate change from all aspects. Based on the available data, there is no direct impact of earthquakes on climate change, which has led to the lack of interest.

R3: Remote Sensing Center Director stated that there is an interest, as previously mentioned in the first answer. Thank God the center is concerned with issues of climate change, and there are numerous research projects in this regard. This is from the research perspective. Additionally, the center organizes conferences in collaboration with international scientific institutions. In 2022, we had an international scientific forum discussing the impact of technology utilization and hurricane mitigation. We have also organized lectures and scientific workshops involving participants from 25 countries around the world.

R4: The Director of Omani Studies Research Center reported that the center has conducted more than six studies focusing on climate change awareness, climate change-related disasters, and their inclusion in the curriculum.

As for the availability of the appropriate infrastructure for conducting research related to climate change Dr. Charabi (former Director of Environmental Studies Research Center), said that they don't have infrastructure for research in climate change. It is based on personal initiatives; we don't have a well-structured track of research on climate change. It is a personal interest, not an institutional decision. This is because people who are running the research are coming from the College of Engineering and for them, everything should be related to engineering. It is not based on institutional thinking; it is based on the personal profile of the person who is running the research. I remember when the person in charge was from the College of Arts, the research focus was on areas related to the humanities, learning and teaching. Now we have people from the College of Engineering, so the situation has changed.

We have theoretical knowledge to assess the level of vulnerability and to develop applied research, because climate change is not only to develop what we call academic papers. We need to develop the technology that is suitable to Oman. We need applied research. We need to develop applied technology that suits the hot climate of Oman, because every country in the world pushes their scientists to develop solutions that could be applied efficiently within their national context. Oman is not doing this and waiting for people outside to do this research for them.

R2: Earthquake Research Centre Director stated that "there is currently no infrastructure available due to the lack of a direct connection between earthquakes and climate change. However, when it comes to seismic monitoring infrastructure, it is well established with observatory devices and monitoring stations." As for infrastructure related to climate change at the Seismic Monitoring Center, it does not exist. However, it would be possible to add a weather station since it can be transmitted through satellites, but the center currently does not have such a station.

R3: Remote Sensing Research Center Director confirmed the availability of infrastructure needed for research. He stated that "the center has all the necessary infrastructure, including the geographic dimension, which enables the analysis of aerial images through various means to address issues and challenges related to climate change. The center also has the latest devices and software in geographic information systems used for image processing, terrain analysis, radar analysis, digital imaging, data imaging, digital transformation devices, color scanning, as well as spectrophotometers for analyzing data related to remote sensing, minerals, soil, crops, marine environment, and water."

R4: According to the Oman Studies Center Director, the center studies and approach climate change from various perspectives: educational, economic, social, and cultural. These aspects do not necessarily require research facilities such as laboratories. Therefore, the center has been consistently conducting these studies.

Regarding the Center's plans, intention or readiness to fund and conduct research on climate change, R1 explained: "Since 2019, we have not been able to generate new scientific knowledge because of the lack of funding. The available funds from the ministry and the government were allocated mainly for software, hiring research assistants, and developing data." He also added that: "Last year we submitted two proposals on climate change for His Majesty's grant program, but were both rejected as the topics were not considered a priority."

At present, the Center's strategy is to seek funding from international organizations because the funding amount at SQU is insufficient for serious and valuable climate change research, as the collection of required data is very expensive. For example, modeling requires specialized expertise, access to supercomputers, and well-equipped laboratories. These resources we currently lack. We also do not have the necessary capacity-buildings programs to support this kind of research.

Meanwhile, international ranking organizations require universities to transition to low-carbon campuses. This means Sultan Qaboos University needs to assess its carbon emissions, including energy used for electricity, and then developing projects to reduce emissions. Institutions are expected to cut their carbon footprint by at least the same amount that they produce. In response, His highness requested two proposals: one for installing 20 megawatts of renewable energy at the university, estimated to cost around 50 million USD. Since neither the government nor the university has the resources to cover this cost, R1 stated that: "I was asked to prepare a proposal for submission to the Green Climate Fund. However, we have been unable to submit this proposal because the university does not produce the required financial transparency reports. For over two years, we have struggled with this issue. Although a Green Campus Committee has been established, the same problems persist, making it extremely difficult to attract funding from international organizations."

According to the Earthquake Centre Director (R2), the Center has no intention of funding research projects due to the nature of its work. However, he suggested that other centers closely associated with climate change may be interested in providing funding opportunities. In contrast, the Remote Sensing Center Director (R3) explained that although their Center is not a funding entity, it can provide research consultations and engage in collaborative research projects funded by external requesting parties, who present specific challenges for which the center develops solutions. Meanwhile, the Oman Studies Centre Director (R4) confirmed that the center actively encourages researchers from Sultan Qaboos University and other institutions to submit research proposals for funding and is also prepared to offer research consultations on climate change to both public and private sector organizations.

Concerning suggestions for enhancing the interest of the research centers at Sultan Qaboos University (SQU) in conducting studies related to climate change, (R1) emphasized that while the university has strong scientists, there is a need to develop structured programs and clear research tracks that addresses both theoretical and applied research. He highlighted the importance of applied research in areas such as water, public health, and climate change technology, we need applied research. The university here is connected to what the government decides. R1 also noted that research priorities at SQU are closely aligned with government directives, for instance, after the adoption of Oman Vision 2040 as a guiding framework, attention quickly shifted when the government later announced a focus on carbon neutrality, promoting the university to align its programs and proposals accordingly.

The analysis of the interviews also revealed four central themes concerning the enhancement of climate change research within SQU research centers. These themes are:

Capacity Building Expertise A strong need was identified for building human capacity to advance climate change research. This includes recruiting specialized researchers, providing continuous professional development, and ensuring staff are equipped with the latest technologies and research skills and methodologies. Interdisciplinary collaborations across centers were also emphasized as a way to integrate expertise from political, economic, social, cultural, and environmental perspectives.

Financial Support and Resource Allocation The availability of financial resources emerged as a critical factor in enabling effective research. Participants highlighted the importance of securing adequate funding for individual projects as well as for long-term strategic research. Sustained financial and institutional support was viewed as essential for addressing the complex and multidimensional challenges posed by climate change.

Research Integration and Strategic Positioning Another key theme was the need to embed climate change within broader institutional and national research agendas. This involves consistently prioritizing studies on the causes and impacts of climate change, as well as promoting applied research focused on mitigation strategies. Integrating climate change into the agendas of research bodies and universities was seen as vital for elevating its strategic importance.

National and Policy-Level Impact Finally, the findings underscored the role of research in shaping national strategies. Climate change research was viewed not only as an academic pursuit but also as a contributor to policy development. Interdisciplinary studies were seen as particularly valuable in informing comprehensive national strategies to respond effectively to the multifaceted challenges of climate change.

10.5 Conclusion

The analysis of collected data revealed that research studies on the issues of climate change and its impact on sustainable development in Oman is insufficient. Research centers in Sultan Qaboos Universities need to adopt comprehensive strategies and develop research projects that directly address this phenomenon and its implications for sustainable socioeconomic development in the Sultanate of Oman. The findings further indicate that climate change research conducted by research centers at SQU is very limited and largely dependent on the personal interest of individual researchers. As a result, the potential role and impact of research findings on developing government policies and guiding practical actions is significantly underutilized.

Enhancing research productivity on climate change requires deliberate and sustained efforts. Key measures include, but not limited to, allocation of research funding, the recruitment of specialized researchers, the establishment of capacity-building initiatives, and the explicit inclusion of climate change as a priority area within the strategic plans of the university and its research centers. Without such measures, the contribution of academic research, conducted by higher education institutions, to climate change policy formulation and sustainable development will remain marginal.

Moreover, the analysis demonstrated that none of the research centers at SQU currently conduct intensive or systematic research on climate change or its related dimensions. Although a few projects have been undertaken, they remain fragmented, based on personal interest, lack continuity, and their overall benefit and impact are limited and not clearly articulated. This gap highlights the urgent need for more comprehensive and coordinated research efforts that can provide evidence-based insights into climate change vulnerability and adaptation in the Sultanate of Oman.

Finally, SQU researchers and research centers concerned with climate change issues must assume a more proactive role in advancing the national research agenda. By strengthening their engagement and producing high-quality, policy-relevant research, they can play a pivotal role in supporting Oman's pursuit of the sustainable development goals and in addressing the pressing challenges posed by climate change.

References

Al-Saidi M, Elagib N (2018) Ecological modernization and responses for a low-carbon future in the Gulf Cooperation Council countries. Climate Change. Wiley Periodicals, Inc. https://doi.org/10.1002/wcc.528

Al-Saraf E (2022) What (exactly) is a research gap? https://gradcoach.com/research-gap. Accessed April 15, 2023

Annual Health Report (2021) Ministry of Health, Sultanate of Oman. https://www.moh.gov.om/

Baako I, Alhassan H, Gidisu P (2022) Understanding and spotting research gaps through a systematic literature review. Int J Res Innov Soc Sci (IJRISS) 6(3):549–554

Bai X, Alyson S, Thomas E, Franz WG, Burak G, Susan P, Anne-Hélène PR et al (2016) Defining and advancing a systems approach for sustainable cities. Curr Opin Environ Sustain 23:69–78

Battaglini M, Harstad B (2016) Participation and duration of environmental agreements. J Polit Econ 124(1):160–204. https://doi.org/10.1086/684478

Cammalleri C, Naumann G, Mentaschi L, Formetta G, Forzieri G, Gosling S, Bisselink B, De Roo A, Feyen L (2020) Global warming and drought impact in the EU, EUR 29956 EN, Publications Office of the European Union, Luxembourg. https://doi.org/10.2760/597045

Charabi Y (2013) Projection of future changes in rainfall and temperature patterns in Oman. J Earth Sci Clim Change 4(5) https://doi.org/10.4172/2157-7617

Fazey I, Hughes C, Schäpke N, Leicester G, Lee E, Goldstein B, Hodgson A, Mason-Jones A, Moser S, Sharpe B, Reed M (2021) Renewing universities in our climate emergency: Stewarding system change and transformation. Front Sustain https://doi.org/10.3389/frsus.2021.677904

Harstad B (2016) The dynamics of climate agreements. J Eur Econ Assoc 14(3):719–752. https://doi.org/10.1111/jeea.12138

Internal Displacement Monitoring Centre (IDMC) (2017) Global report on internal displacement, grid

Jones D (2012) Looking through the "greenwashing glass cage" of the green league table towards the sustainability challenge for UK universities. J Organ Chang Manag 25(4):630–647

Miles D (2017) A taxonomy of research gaps: identifying and defining the seven research gaps, Doctoral Student Workshop: Finding Research Gaps—Research Methods and Strategies, Dallas, Texas

Müller-Bloch C, Kranz J (2014) A framework for Rigorously identifying research gaps in qualitative literature reviews. In: Proceedings of the 36th international conference on information systems, Fort Worth, pp 1–19

Qureshi F (2019) Don't know where to start? 6 Tips on identifying research gaps. Editage Insights. https://www.editage.com

Robert K, Parris T, Leiserowitz A (2005) What is sustainable development? goals, indicators, values, and practice. Environ Sci Policy Sustain Develop 47(3):8–21. https://doi.org/10.1080/00139157.2005.10524444

Robinson K, Saldanha I, McKoy N (2011) Development of a framework for to identify research gaps systematic reviews. J Epidemiol 64(1):1325–1330

Salas E (2024) Number of hurricanes globally 1990–2023. Statista. https://www.statista.com/statistics/1297656/number-hurricanes-worldwide/

Schoch-Spana M, Gill K, Hosangadi D, Slemp C, Burhans R, Zeis J, Carbone E, Links J (2019) Top-down and bottom-up measurement to enhance community

resilience to disasters. Am J Public Health 265–267. https://doi.org/10.2105/AJPH.2019.305151
Stein S, Andreotti V, Suša R (2019) 'Beyond 2015', within the modern/colonial global imaginary? global development and higher education. Critic Stud Educ 60(3):281–301. https://doi.org/10.1080/17508487
Sustainable Development Solutions Network (SDSN) (2017) Getting started with the SDGs in universities. https://resources.unsdsn.org/getting-started-with-the-sdgs-in-universities.
Symanski E, Heyreoun AH, Inkyu H, Michelle M, Kristina WW, Sheryl M, William BP et al (2022) Responding to natural and industrial disasters: partnerships and lessons learned. Disast Med Public Health Prepared 16(3):885–888. https://doi.org/10.1017/dmp.2020.467
Toreti A, Bavera D, Juan A, Carmelo Cammalleri C, Jager A, Ciollo C, Essenfelder A, Maetens W, Magni D, Masante D, Mazzeschi M, Niemeyer S, Spinoni J (2022) Drought in Europe August 2022. Publications Office of the European Union, Luxembourg, https://doi.org/10.2760/264241
Yamin F, Joanna D (2004) "Contents." Toc. In: The international climate change regime: a guide to rules, institutions and procedures. Cambridge University Press, Cambridge, pp v–xiv

Omer Hamad is an Associate Professor of Educational Administration at the College of Education, Department of Educational Foundations and Administration in Sultan Qaboos University, Sultanate of Oman. Dr. Omer graduated with a Ph.D. degree in Educational Administration from Pennsylvania State University (USA). His areas of research interest include educational policies and planning strategies, teacher education; implementation of child rights laws; educational change and politics of educational decision making, and human resources development. Dr. Omer has a rich work experience in higher education institutions in the Sudan, Arab Gulf states, and USA.

11 The SDGs and Universities: Promoting Innovation and "Living Together" in the UAE

Salem Alhammadi and Abdeljalil Akkari

Abstract

This chapter investigates the engagement of four universities in the United Arab Emirates with the SDGs. Focusing on the integration of Education for Sustainable Development, the study examines how these institutions contribute to sustainable development. It analyzes the universities' participation in addressing social challenges at national, regional, and global levels, assessing their visibility and reporting of SDG initiatives online. Findings show significant differences in SDG engagement and reporting across the universities. The study underscores the need for enhanced SDG reporting and visibility within the UAE's higher education sector. It concludes with recommendations for future research, emphasizing the importance of expanding the scope of study and integrating diverse methodologies for a more comprehensive understanding of universities' roles in advancing the SDGs.

Keywords

Sustainable development goals · Global citizenship education · Higher education · Universities · United Arab Emirates · SDG reporting

11.1 Introduction

By addressing the various Sustainable Development Goals (SDGs) issues in teaching, research, and service, universities can support staff and students to become agents of positive change for sustainability and prepare themselves to apply these skills for a (more) sustainable world (Mori Junior et al. 2019). In particular, the introduction of Education for Sustainable Development (ESD) in the 2030 Sustainable Development Agenda has influenced the political agenda of national governments worldwide to integrate sustainability in education policy and practice. Universities across the world are currently helping to meet the SDGs in a number of areas. In addition to research and teaching, these also include activities related to knowledge transfer, operations management, and engagement as public institutions (Leal Filho et al. 2021). For instance, the UAE University publishes detailed reports covering all 17 SDGs, while the University of Sharjah also provides

S. Alhammadi (✉) · A. Akkari
University of Geneva, Geneva, Switzerland
e-mail: Me.Salem3@gmail.com

A. Akkari
e-mail: abdeljalil.akkari@unige.ch

K. Al'Abri and E. Buckner (eds.), *Higher Education and Sustainable Development in the Arab Gulf States*, Sustainable Development Goals Series, https://doi.org/10.1007/978-3-032-19559-3_11

a clear, data-supported report. In contrast, Zayed University and the Higher Colleges of Technology lack comprehensive SDG reporting, with some goals missing or only partially addressed. This variation highlights the uneven visibility and integration of SDGs in UAE higher education.

Living together refers to the context of diverse groups and individuals who come from different cultures, speak different languages, and follow different religions, yet coexist within a single society under one legal framework. It emphasizes their capacity to interact with one another on a daily basis in ways oriented toward minimizing conflict.

This chapter examines how universities in the United Arab Emirates engage with the SDGs, with particular attention to the integration of ESD within their institutional frameworks. It seeks to shed light on the roles and strategies of UAE universities in advancing the SDGs, especially through the themes of innovation, and the principle of *living together*. The notion of *living together* is especially pertinent in the UAE's social context, marked by its cultural, linguistic, and demographic diversity.

These concepts are interconnected in the way they contribute to the main goal of sustainable development. UAE universities, by including SDGs in their curriculum and fostering a culture of innovation, play a crucial role in shaping students who are not only knowledgeable in their fields of study but are also conscious global citizens committed to contributing to a sustainable future.

The UAE was established in 1971 as a federation of seven emirates: Abu Dhabi, Dubai, Sharjah, Ajman, Umm Al Quwain, Ras Al Khaimah, and Fujairah. It is a member of the Gulf Cooperation Council (GCC), which includes Bahrain, Kuwait, Oman, Qatar, and Saudi Arabia. Similar to the other GCC countries, the UAE suffers from a substantial shortage of qualified local workers and relies heavily on foreigners to fill the gap. More than 80% of the UAE labor workforce are expatriates of more than 100 nationalities, reflective of many races, cultures, ethnicities, and languages (Ababneh and Al-Waqfi 2016; Baruch and Forstenlechner 2017).

The case of the UAE illustrates how global interdependencies shape local societies. With citizens making up <11.5% of the total population, the country depends heavily on a large and diverse expatriate community, making questions of integration especially significant. In response to their growing numerical *minority status*, many GCC states, including the UAE, have become more stringent about their citizenship, nationality, and employment policies (Jamal 2015; Leimgruber et al. 2022). New citizenship policies that some assumed would accompany globalization may be an opportunity to promote a "*living together*" policy in multicultural societies particularly in school context (Alabri et al. 2022; Moon and Tocci 2020). The development of the UAE higher education system demonstrates evidence of a generally positive attitude toward westernized education systems in the UAE, including the use of English-medium education. This choice by the UAE government appears to be based on the principle of modernization, the recognition of higher education as a nation-building tool and out of respect for the achievements of the Western education model (Burden-Leahy 2009). In sum, although concerns about the UAE's demographic structure have reinforced an emphasis on national identity, the country simultaneously recognizes the realities of global interdependence. This tension is particularly visible in higher education, where universities are positioned both as tools for nation-building and identity formation, and as globally oriented institutions that adopt English as the medium of instruction and engage with international agendas such as the SDGs.

In this chapter, our analysis consists of two parts. In the first we will analyze the initiatives taken in four major universities in the UAE to promote the SDGs. In particular, we investigate how committed universities are finding solutions to social challenges on the national, regional, and global level, such as those defined in the United Nations SDGs. We also analyze the web visibility of the SDGs and the 2030 Agenda on UAE university websites (Vallez et al. 2022).

In the second part, we focus on the status and definition of GCED in university programs in the UAE. As a paradigm, GCED "exemplifies how education is able to develop the knowledge, skills, values and attitudes that learners need to achieve a more just, peaceful, tolerant, inclusive, safe and sustainable world" (UNESCO 2014). By combining the cognitive, socio-emotional, and behavioral dimensions of learning, GCED applies a holistic approach adaptable to different contexts. We will evaluate if GCED is appropriate in the context of the UAE as the question of national identity is quite sensitive.

This chapter seeks to bridge the knowledge gap in understanding how universities in the United Arab Emirates advance the SDGs. It examines their contributions to national, regional, and global sustainability challenges, as well as their strategies for promoting SDG-oriented education. Drawing on a mixed-methods approach that combines web visibility analysis with document review, the chapter critically evaluates the scope and impact of these universities' SDG initiatives. The central argument is that, although UAE universities are increasingly engaging with the SDGs, the degree of integration and visibility remains uneven, highlighting the need for a more systematic and comprehensive approach to embedding the SDGs within institutional operations and curricula.

11.2 Literature Review

As the theme of the SDGs is relatively new for universities, it seems useful to us to synthesize international research. The literature review is subdivided into three parts. In the first, we will analyze the results of research on the presence of the SDGs at the university. In the second part, we focus on identifying research on the SDGs in the UAE. In the third, we will discuss research on the concepts of *living together* and *global citizenship* in higher education in the Arab world.

Universities are uniquely placed to lead the cross-sectoral implementation of the SDGs by providing an invaluable source of expertise in research and teaching. Their role in this domain is diverse, and existing studies show mixed results. Leal Filho et al. (2019) argue that implementation of the SDGs at universities remains in its infancy, though if embedded effectively into teaching and learning, SDGs could strengthen universities' relationships with external stakeholders and society. Mawonde and Togo (2019) provide a concrete example from South Africa, identifying four main types of initiatives: operations and management, community engagement, teaching and learning, and research. Taken together, these studies suggest that while universities recognize the relevance of SDGs, most institutions are still experimenting with relatively small-scale initiatives, indicating a gap between broad conceptual advocacy and systematic institutional practice.

One of the most challenging tasks universities face is raising awareness about the SDGs. Because the agenda is still relatively new, there remains a lack of widespread knowledge about the 17 goals, their indicators, and their targets. This challenge is particularly important to address given the SDGs' guiding principle that "no one must be left behind" (Mori Junior et al. 2019). Evidence from Spain illustrates the scale of the problem: knowledge of SDGs among university students is very limited, mirroring the general lack of awareness among the wider Spanish population. Although some initiatives have been launched to promote SDG awareness at the university level, they have not yet been generalized across institutions (Zamora-Polo et al. 2019). Together, these findings indicate that low levels of awareness are not only a matter of student knowledge but also of institutional dissemination strategies, highlighting the need for universities to move beyond isolated initiatives toward more systematic approaches.

The institutionalization of partnerships around the Sustainable Development Goals (SDGs) with government and community actors represents a central strategy for advancing sustainable development. Such institutionalization serves multiple and interrelated functions within higher education and policy ecosystems. It strengthens the science–policy interface by fostering evidence-informed decision-making and ensuring that academic research contributes directly to public policy and practice. It also supports the development of integrated and coherent policies and actions that align with the multidimensional nature of the SDGs. Through these partnerships, universities become more actively involved in political processes, contributing expertise and critical perspectives to national and local development agendas. Moreover, institutionalized collaboration enhances transdisciplinary learning and educational interactions, enabling knowledge exchange across sectors and disciplines. Finally, these initiatives demonstrate a visible and sustained commitment to effective engagement and societal impact, reinforcing the role of higher education institutions as key actors in achieving sustainable development (El-Jardali et al. 2018).

Another proposed strategy is to conceptualize universities as *Agents of Change* (AOC). The traditional role of the university as primarily a knowledge institution has evolved into a broader, outward-facing mission. Collaboration with external partners is increasingly standard practice, allowing the human capital, research, and expertise produced by universities to generate greater societal impact. In this view, the university acts as an AOC in its immediate community by fostering mutually beneficial relationships, while consistent knowledge transfer across institutions creates opportunities for global influence (Brugmann et al. 2019). This perspective complements other frameworks such as "Living Labs," as both emphasize universities' active engagement beyond campus boundaries; however, the AOC approach places more weight on sustained institutional partnerships, whereas the Living Lab model highlights experiential and student-centered learning.

Another development in HEIs is the growing emphasis on experiential learning to address pressing societal issues. This has often taken the form of solutions-based pedagogy, guided by collaboration with partners outside academia such as university staff, civil society organizations, and private sector actors. One widely cited example is the *Living Lab* approach at the University of Toronto, also referred to as "real-world laboratories," "urban Living Labs," or "sustainability learning labs." Institutions employing this model share consistent principles, including interdisciplinarity, collaboration, and direct student engagement in sustainability challenges (Brugmann et al. 2019). The University of Toronto's program illustrates this through a three-tiered pathway: *Sustainability Citizen*, which encourages participation in extracurricular sustainability activities; *Sustainability Scholar*, which integrates sustainability into credit-bearing courses with certification; and *Sustainability Leader*, a comprehensive pathway combining curricular and co-curricular engagement, international experience, and a capstone project. These initiatives demonstrate how Living Labs can institutionalize sustainability across multiple levels of student engagement. Compared with the *Agent of Change* framework, which emphasizes institutional partnerships, the *Living Lab* approach highlights the pedagogical dimension by embedding sustainability into the student experience. However, while the model has shown promise in North American universities, its transferability to different contexts—such as the UAE, where higher education is strongly linked to national policy goals—requires further investigation.

Universities occupy a unique position within society. With a broad remit around the creation and dissemination of knowledge, universities have long been powerful drivers of global, national, and local innovation; economic development; and societal well-being. As such, universities have a critical role in the achievement of the SDGs and will also greatly benefit from engaging with them (Purcell et al. 2019; Kestin et al. 2017).

In 2017, the UAE's Ministry of Education (MOE) launched its National Strategy for Higher Education 2030, prioritizing quality higher education, particularly with respect to research and the country's focus on becoming a knowledge economy. The 2030 plan is focused on developing students, linking academia and the labor market, engaging the private sector in curricular development, promoting and expanding the production of research, and increasing the global competitiveness of its higher education institutions by creating new and innovative academic programs. This expansion is aimed specifically at producing specialized and professional graduates who can compete in vital global sectors by producing research, engaging in entrepreneurship and joining the labor market (Ashour 2020). The strategy reflects broader international trends positioning higher education as a driver of knowledge economies, but in the UAE context it also raises questions about balancing global competitiveness with local identity and social needs.

UAE universities and the SDGs engage in a mutual relationship, where academic institutions leverage their research and innovation to address key challenges highlighted by the SDGs, such as those identified by El-Zein et al. (2016) in the context of health and environmental sustainability. This alignment with the SDGs enriches the universities' educational and societal contributions, positioning them as pivotal agents in the UAE's strategy for sustainable development as outlined in the National Strategy for Higher Education 2030. This role resonates with the earlier notion of universities as Agents of Change (AOC), since UAE institutions are increasingly expected not only to generate knowledge but also to translate it into tangible social and environmental outcomes through sustained partnerships and innovation.

In the UAE, Gulseven and Ahmed (2022) analyzed the indicators that measure the UAE's performance toward achieving SDG 15. This research sets a strategy for preserving biodiversity in the UAE and attaining the relevant goals to protect life on land. The findings suggest that while the UAE is motivated to achieve the goals stated in SDG15, significant challenges remain ahead. Specifically, many of the relevant indicators are missing, making it hard to measure country performance based on reliable metrics.

El-Zein et al. (2016) studied the challenges faced by the Arab world in achieving the SDGs, noting the region's struggle with poverty, poor health, and significant environmental degradation. The study highlighted persistent infectious diseases in lower to middle-income Arab nations and increasing non-communicable diseases across all Arab countries. Other critical issues include high personal health expenditure, limited access to safe water, violent conflicts, foreign interventions, and severe social and political fragmentation leading to weak health systems and restricted health rights. The research pointed out two extreme indicators in the region: a refugee per capita ratio ten times the global average and the highest percentage of international migrant workers in the world, especially in the Gulf Cooperation Council countries. Additionally, the region has a high level of militarization with per capita weapons imports four times higher than the global average, coupled with the lowest health-to-military expenditure ratio, which is less than a fifth of the global average. This high militarization not only diverts resources away from poverty reduction and sustainable development but also raises questions about the effectiveness of international aid, as financial outflows for arms in some Arab nations surpass the assistance received from wealthier countries. While the authors stress the critical role of the state in addressing these issues, their findings also point to the importance of universities as potential contributors to solutions, whether through health research, policy analysis, or training future leaders capable of navigating the interconnected challenges of migration, conflict, and sustainable development.

Nonetheless, there are numerous challenges confronting educational governance in the UAE's higher education sector (Jose and Chacko 2017). These include inconsistent academic requirements, diverse employer recognition of academic credentials, and the complex landscape of online education accreditation, compounded by stringent review processes. Together, these issues affect the efficiency and credibility of the system, with direct implications for its contribution to economic and social sustainability. Addressing such governance challenges is therefore not only a matter of administrative reform but also central to enabling universities to fulfill their broader role in advancing the SDGs. In this sense, weaknesses in governance risk undermining universities' ability to act as *Agents of Change* and to integrate innovative models such as Living Labs into their teaching and research practices. Governance challenges in higher education are not only technical or administrative but also deeply connected to broader social questions in the region. In particular, the issues of "living together" and citizenship are central to analyzing the future of the MENA region. Waghid and Davids (2014) highlight the (im)possibility of democratic citizenship education in this context, underscoring the tension between national identity and the demands of inclusive, globally connected education.

El-Zein et al. (2016) highlight critical issues surrounding citizenship, labor migration, and refugee flows in the Arab world, in the context of Sustainable Development Goals (SDGs). The region is marked by a significant presence of non-citizens, including refugees and migrant labor, which underscores the relevance of discussing citizenship rights. In fact, the GCC countries are home to a large non-citizen population, constituting about 49% of the total population. Arab nations that are not oil producers send out a substantial number of labor emigrants, with remittances playing a crucial role in their GDPs. However, working conditions are often poor and characterized by ethnic discrimination, with migrant workers exposed to exploitation and abuse, especially under the Kafala system. Non-citizens, including those born in the region, struggle to gain citizenship and have limited access to basic rights. Simultaneously, the Arab world hosts a significant number of refugees, with countries like Lebanon and Jordan having refugees constituting up to 38% of their populations. Much of these refugee movements are due to continuous conflicts and unrest in many Arab countries over the past 25 years, often involving foreign intervention.

In summary, the literature highlights the SDGs as an opportunity to rethink higher education. In the UAE, this opportunity is all the more important as the country has built its prosperity on exhaustible fossil fuels and on the contribution of international migrants bringing an ethnic diversity unmatched elsewhere in the world.

The literature review has revealed a gap within the already existed academic work regarding the topic of this chapter, which is mainly related to the UAE context. The gaps are identified on the following areas:

The first gap concerns the lack of focus on UAE universities. While existing literature extensively discusses higher education's role in SDGs, there is a scarcity in studies related to the UAE. This chapter, therefore, presents an examination of how UAE universities approach SDGs. The second gap involves the integration of innovative teaching strategies for SDGs in the UAE. Although the subject of teaching SDGs is not new in academic literature, the specific exploration of innovative pedagogical approaches within the UAE context remains underexplored. The third gap is the lack of detailed case studies or practical examples from UAE universities. This chapter addresses this research gap by offering concrete examples and case studies, showcasing practical implementations of SDGs in the UAE's higher education sector, by showcasing four Emirati universities. Finally, the fourth gap is the combined focus on GCED and Education for ESD within the UAE's educational institutions. This chapter contributes to this underrepresented area by illustrating how UAE universities operationalize GCED and ESD within their curricula and institutional strategies.

11.3 Data and Methods

Our chapter was informed by a mixed methods research design and was based on a case study of universities in the UAE. Data collection was based on qualitative techniques including document analysis and websites observations. According to Creswell (2003), qualitative research is a means for exploring and understanding the meaning individuals or groups assign to a social human problem. The process of research involves emerging questions and procedures, analyzing the data inductively and making interpretations of the meaning of data. As suggested by Yin (2003), data for case studies can come from six sources of evidence including documentation. We used university websites as a source of documentation in this research.

This study's objective is to assess UAE universities' engagement with Sustainable Development Goals, particularly focusing on Global Citizenship Education and Education for Sustainable Development. It aims to analyze how these institutions address societal challenges and their strategies in communicating SDG initiatives, especially through web visibility and reporting.

Our research employed a mixed method, utilizing document analysis and website observation, as they are the only available resources for this topic. This approach aimed to validate the information and provide detailed insights into the contributions of universities in the UAE toward the SDGs. The primary sources for our study were the official websites of each university and their published reports. Our process involved an initial examination of each university's website to collect information related to the SDGs, followed by an analysis of the reports to complete details not covered by the website.

For the analysis, we employed a qualitative content analysis approach. Documents and website materials were systematically examined using a coding strategy adapted from Braun and Clarke's (2006) thematic analysis framework. Initial codes were generated inductively from the data (e.g., presence or absence of SDG references, number of goals explicitly mentioned, type of initiative described) and were then organized into broader themes such as teaching, research, community engagement, and operations.

To evaluate web visibility, we developed a set of criteria including (a) whether SDG-related content was directly accessible from the university's homepage, (b) whether official reports on SDGs were published, and (c) whether the 17 goals were explicitly represented or only partially covered. Each university was assessed against these criteria. Although the study was limited to a single coder, a consistent coding protocol was applied across all cases to ensure comparability. Future research could strengthen reliability by incorporating multiple coders and calculating intercoder agreement.

The methodology we used possesses certain limitations that need to be acknowledged. The main constraint is the limited sources of information, as our data collection is limited to document analysis and website observations. Another important limitation is the scope of our study. We focused mainly on the four largest universities in the UAE in terms of student numbers. However, there are 74 universities in the UAE in total (Higher Education Institution 2020). Our findings may not be representative of the broader landscape of higher education in the country. Furthermore, our methodology did not include direct inputs from various stakeholders within these universities. Another limitation is that relying solely on websites and published reports may not fully capture the extent of universities' actual activities. Online documents often reflect official narratives and institutional self-presentation, which may differ from on-the-ground practices and stakeholder experiences. As such, our findings should be understood as indicative of how universities choose to communicate their SDG engagement, rather than a comprehensive account of all activities undertaken.

The universities selected for this study are the United Arab Emirates University (UAEU), Zayed University (ZU), Higher Colleges of Technology (HCT), and University of Sharjah (UoS). While financial funding details are not available, the first three are public-federal universities, while the University of Sharjah is a public-local university. UAEU hosts 14,900 students from 82 countries and is recognized as the first university in the country (UAEU 2024). HCT, with its 21,500 students, operates across

16 campuses nationwide, primarily focusing on applied sciences (HCT 2021). Zayed University, with a total of 11,153 students across its campuses in Abu Dhabi and Dubai, is known for its high ratio of female to male students (ZU 2023). In the case of UoS, it is the only public-local university in our list, with a total student population of 19,240 (UoS n.d.).

The web visibility analysis in our study involves assessing whether information related to the SDGs of each university appears on search engines like Google and Bing. These platforms have become essential for sourcing content on the internet. The significance of this approach lies in providing an overview of a university's outreach efforts regarding its SDG initiatives (Vallez et al. 2022). Our methodology comprises a two-step process. The first step entails directly examining the university's website for SDG content. If no results are found, the second step involves using Google with the search keywords "University name" and "SDGs." This uniform set of keywords was applied across all universities.

To address the limitations of our current study, future research should incorporate additional methods such as interviews, surveys, and syllabus analysis. Interviews could be conducted with university top-level personnel, heads of colleges, or staff to gain insights from different administrative levels. Surveys targeting students would provide perspectives from the student body. Moreover, analyzing the content of syllabi would offer a deeper understanding of the specific SDG-related topics being taught in classes. Furthermore, analyzing the content of SDGs within a certain framework is necessary to examine its value, reliability with the goal, and its impact.

11.4 Findings

The UAE University's work on the SDGs is very visible on its website, which features its SDG efforts and showcases a detailed report that outlines activities across all 17 goals, complemented by quantitative data to illustrate its progress. The university's contributions, especially in quality education and partnerships to achieve the goals, are marked by various initiatives and collaborations. However, UAEU's display of SDGs has dual aspects. The first is an online display, which shows the SDG content on the website page. The second aspect is a report which includes detailed information and data (Sustainable Development Goals 2023).

In

ZU's case, although its SDGs content appears in search engine results, the webpage itself is not well organized and misses two of the 17 goals. The information presented often lacks direct connection to the SDGs, offering general facts rather than specific university-related data. Furthermore, ZU has not provided a report that encapsulates its total contribution to the SDGs, nor has it offered quantitative data showing annual progress (ZU 2023).

For the HCT, while SDG-related content appears in search engine results, the website mentions only 6 out of the 17 goals, often lacking specific linkage to the SDGs and providing general initiatives without direct impact. Additionally, there is an absence of quantitative data or a comprehensive report detailing their contributions to these goals (HCT 2021).

In the case of the UoS, the SDGs report is easily found via the search engine. The report is well organized and shows the details of each goal alone. The report indicates what are the actions taken in each of the 17 goals; they supported their claims with graphs and numbers. Compared to the rest of the universities, Sharjah University has a very clear report (UoS 2022).

Comparing the four universities, it becomes apparent that Zayed University and HCT are lacking in detailed information about their SDG initiatives, with many of their mentioned efforts either having no direct link to the SDGs or being conducted outside their capacity. In contrast, the University of Sharjah and UAEU provides a clear report on its SDG-related efforts. The university has successfully participated in local initiatives, such as providing health assistance during the COVID-19 pandemic, partnering with local entities to build sustainable projects, and raising awareness about sustainability within the local community.

While all four universities demonstrate web visibility in Google search results, the content quality varies significantly. Universities with dedicated reports tend to present well-structured, data-driven, and quantitative information. In contrast, those relying solely on webpages for displaying SDG data often feature weaker content, characterized by a lack of detailed data, irrelevant information, and an absence of quantitative analysis (Table 11.1).

11.5 Discussion

Our analysis of four UAE universities revealed significant differences in their engagement with the SDGs. Zayed University and the Higher Colleges of Technology, both federal institutions, showed limited SDG reporting, relying mainly on fragmented website mentions rather than comprehensive reports. In contrast, the University of Sharjah, a local university, displayed a more systematic approach, including its response to the COVID-19 pandemic and environmental issues. These findings confirm the inadequacy of current SDG reporting practices across UAE universities, and highlight the need for a more comprehensive, standardized approach.

The absence of reporting that addresses all 17 SDGs, and the lack of accessible, up-to-date information, underscores the challenge of ensuring visibility and accountability. While some universities make efforts in selected areas, there remains no holistic strategy that integrates the SDGs across teaching, research, community engagement, and operations. This gap reflects broader governance challenges in UAE higher education, as discussed earlier, where institutional practices are not always aligned with the expectations of the Ministry of Education. Although the UAE's National Strategy for Higher Education 2030 emphasizes research, innovation, and global competitiveness, it does not explicitly require universities to institutionalize SDG reporting. As a result, universities treat SDG engagement as optional or ad hoc rather than embedding it as a structural priority.

Viewed through the frameworks highlighted in the literature review, universities in the UAE are not fully realizing their potential as *Agents of Change*. While the AOC approach stresses sustained partnerships and outward engagement, our findings show that reporting often remains inward-facing and limited in scope. Similarly, although the *Living Lab* model emphasizes integration of experiential learning and sustainability into curricula, few universities in the UAE have translated this into systematic reporting that reflects their pedagogical practices. These inconsistencies indicate that universities' contributions to the SDGs remain fragmented, shaped more by immediate institutional priorities than by long-term national and global goals.

The study findings also lead to concrete recommendations for practice. First, there is a need to develop and implement a standardized SDG reporting framework. This would ensure that all universities track and report their contributions consistently across the 17 goals. Second, there is a need for issuing annual SDG progress reports. Regular, public reports would enhance transparency and comparability between universities, allowing stakeholders to monitor progress over time. Third, it is important to strengthen inter-university collaboration. Universities should exchange practices and experiences related to SDG engagement, fostering collective learning and benchmarking. Fourth, there is a need to better align SDG reporting with the UAE National Strategy for Higher Education 2030. Clearer integration would position SDG initiatives as part of the national agenda on innovation and competitiveness, rather than as voluntary institutional activities. Overall, the findings demonstrate that universities in the UAE play an increasingly visible role in contributing to sustainable development, but that role is constrained by the absence of systematic reporting. To move from fragmented efforts toward meaningful impact, universities must embed the SDGs in their structures and strategies, reflecting both their function as nation-building institutions and their responsibility as global actors.

Table 11.1 Comparison of SDG goals addressed by four major universities in the UAE

SDG goal	UAE University	Zayed University	Higher colleges of technology	University of Sharjah
SDG 1: no poverty	Financial aid programs for students	Not publicly reported	Not publicly reported	Subsidized meals and support funds
SDG 2: zero hunger	Research on food security and sustainable agriculture	Partnership with International Center for Biosaline Agriculture	Not publicly reported	Efforts to reduce food waste and enhance food safety
SDG 3: good health and well-being	Initiatives on public health and well-being	Not publicly reported	Not publicly reported	Health-related community outreach and student wellness programs
SDG 4: quality education	Integration of SDGs in curriculum and research	Not publicly reported	Applied sciences programs linked to SDGs	Curricular and co-curricular sustainability pathways
SDG 5: gender equality	Initiatives supporting women in STEM	Female-majority enrollment highlighted	Not publicly reported	Gender inclusion programs and scholarships
SDG 6: clean water and sanitation	Research on water sustainability	Not publicly reported	Not publicly reported	Projects on water management and conservation
SDG 7: affordable and clean energy	Renewable energy research centers	Not publicly reported	Not publicly reported	Solar energy and sustainable technology projects
SDG 8: decent work and economic growth	Entrepreneurship and innovation hubs	Not publicly reported	Applied vocational training programs	Business incubators and industry partnerships
SDG 9: industry, innovation and infrastructure	Collaboration with industry on R&D	Not publicly reported	Technical and engineering initiatives	Innovation and research clusters
SDG 10: reduced inequalities	Programs supporting students with disabilities	Not publicly reported	Not publicly reported	Inclusion initiatives for marginalized groups
SDG 11: sustainable cities and communities	Urban sustainability projects	Not publicly reported	Not publicly reported	Sustainable campus and community projects
SDG 12: responsible consumption and production	Awareness campaigns on sustainability	Not publicly reported	Not publicly reported	Food waste reduction and recycling programs
SDG 13: climate action	Climate research initiatives	Not publicly reported	Not publicly reported	Environmental awareness campaigns
SDG 14: life below water	Marine biodiversity projects	Not publicly reported	Not publicly reported	Coastal and marine research programs
SDG 15: life on land	Biodiversity and conservation projects	Not publicly reported	Not publicly reported	Reforestation and land protection initiatives

(continued)

Table 11.1 (continued)

SDG goal	UAE University	Zayed University	Higher colleges of technology	University of Sharjah
SDG 16: peace, justice and strong institutions	Governance and policy research	Not publicly reported	Not publicly reported	Legal studies and civic engagement projects
SDG 17: partnerships for the goals	International research partnerships	Not publicly reported	Institutional collaborations with government and industry	Global cooperation projects and networks

11.6 Conclusion

This study evaluated SDG engagement at four UAE universities, UAEU, Zayed University, Higher Colleges of Technology, and the University of Sharjah, and uncovered notable differences in how these institutions report and engage with SDGs. While federal universities showed limited reporting, often relying on fragmented online mentions, the University of Sharjah provided more comprehensive documentation, though still uneven in coverage. These findings highlight the continuing need for improved SDG reporting and visibility across UAE universities.

Beyond documenting current practices, the study points to the broader trajectory of SDG integration in UAE higher education. Universities are uniquely positioned to act as *Agents of Change*, but their impact will remain constrained unless SDG initiatives are systematically embedded into curricula, research, and community engagement. Linking reporting more explicitly to the UAE's National Strategy for Higher Education 2030 could provide the structural incentives currently missing and align institutional practices with national priorities.

An additional dimension emerging from this study is the federal structure of the UAE higher education system. Three of the universities examined, UAEU, Zayed University, and HCT, are federal institutions, while the University of Sharjah operates under local governance. This distinction may partly explain differences in SDG engagement and reporting: federal universities are more closely aligned with national-level strategies, whereas local universities may exercise greater flexibility in shaping their reporting practices. Although our study focused primarily on reporting visibility, a more detailed examination of locally governed universities could reveal further variation and provide richer insights into how governance structures shape institutional approaches to the SDGs.

Moreover, the themes of innovation and living together, introduced at the outset of this chapter, offer critical lenses for interpreting the results. Innovation is essential for designing new frameworks of accountability, enabling universities to move beyond ad hoc reporting toward standardized practices that can be compared internationally. Meanwhile, the concept of *living together* highlights the social responsibility of UAE universities in a highly diverse society. By embedding SDGs into everyday teaching and engagement, universities can contribute not only to sustainable development but also to fostering peaceful coexistence across cultural, linguistic, and religious differences. Practical applications suggested by this study include: developing a unified SDG reporting framework that ensures comparability across institutions; issuing annual SDG progress reports to enhance transparency and allow stakeholders to monitor progress; strengthening inter-university collaboration to foster collective learning and exchange of practices and expanding future research to include interviews, surveys, and syllabus analysis, offering a more comprehensive picture of universities' contributions.

Finally, the integration of SDGs in UAE universities should be seen not merely as a compliance exercise but as a transformative opportunity. If universities succeed in aligning their educational, research, and community missions with the SDGs, they will simultaneously strengthen their role as nation-building institutions and as global actors advancing sustainable development and intercultural coexistence.

References

Alabri KM, Ambusaidi AK, Alhadi BR (2022) Promoting global citizenship education (GCED) in the sultanate of Oman: an analysis of national policies. Sustainability 14(12):7140

Ashour S (2020) Quality higher education is the foundation of a knowledge society: where does the UAE stand? Qual Higher Educ 26(2):209–223

Brugmann R, Côté N, Postma N, Shaw EA, Pal D, Robinson JB (2019) Expanding student engagement in sustainability: using SDG-and CEL-focused inventories to transform curriculum at the University of Toronto. Sustainability 11(2):530

Burden-Leahy SM (2009) Globalisation and education in the postcolonial world: the conundrum of the higher education system of the United Arab Emirates. Comp Educ 45(4):525–544

Creswell JW (2003) Research design: qualitative, quantitative, and mixed methods approaches. Sage Publications, New York

El-Jardali F, Ataya N, Fadlallah R (2018) Changing roles of universities in the era of SDGs: rising up to the global challenge through institutionalising partnerships with governments and communities. Health Res Policy Syst 16:1

El-Zein A, DeJong J, Fargues P, Salti N, Hanieh A, Lackner H (2016) Who's been left behind? Why sustainable development goals fail the Arab world. Lancet 388(10040):207–210

Gulseven O, Ahmed G (2022) The state of life on land (SDG 15) in the United Arab Emirates. Int J Soc Ecol Sustain Develop 13(1):1–15

Higher Colleges of Technology (2021) HCT fact book 2020–2021. https://hct.ac.ae/wp-content/uploads/2021/03/HCT_FACT_BOOK_2020-21_14mar.pdf

Higher Education Institutions - CAA (2020) https://www.caa.ae/Pages/Institutes/All.aspx

Jamal MA (2015) The "tiering" of citizenship and residency and the "hierarchization" of migrant communities: the United Arab Emirates in historical context. Int Migrat Rev 49(3):601–632

Jose S, Chacko J (2017) Building a sustainable higher education sector in the UAE. Int J Educ Manag 31:752–765

Kestin T, van den Belt M, Denby L, Ross K, Thwaites J, Hawkes M (2017) Getting started with the SDGs in universities: a guide for universities, higher education institutions, and the academic sector. United Nations

Leal Filho W, Shiel C, Paco A, Mifsud M, Ávila L, Brandli L, Molthan-Hill P, Pace P, Azeiteiro U, Vargas V, Caeiro S (2019) Sustainable development goals and sustainability teaching at universities: falling behind or getting ahead of the pack? J Clean Product 232:285–294

Leal Filho W, Frankenberger F, Salvia AL, Azeiteiro U, Alves F, Castro P et al (2021) A framework for the implementation of the sustainable development goals in university programmes. J Clean Product 299:126915

Leimgruber JR, Al-Issa A, Lorenz E, Siemund P (2022) Managing and investing in hybrid identities in the globalized United Arab Emirates. J Lang Ident Educ 21(1):1–18

Mawonde A, Togo M (2019) Implementation of SDGs at the University of South Africa. Int J Sustain High Educ 20:932–950

Moon S, Tocci C (2020) Citizenship education beyond the nation-state: implications for teacher education. Global citizenship education and teacher education. Routledge, New York, pp 85–101

Mori Junior R, Fien J, Horne R (2019) Implementing the UN SDGs in universities: challenges, opportunities, and lessons learned. Sustain J Rec 12(2):129–133

Purcell WM, Henriksen H, Spengler JD (2019) Universities as the engine of transformational sustainability toward delivering the sustainable development goals: "living labs" for sustainability. Int J Sustain High Educ 20(8):1343–1357

Sustainable Development Goals (SDGs) (2023) UAEU. https://www.uaeu.ac.ae/en/about/sustainable-development-goals/

UNESCO (2014) Global citizenship education: preparing learners for the challenges of the 21st century. http://unesdoc.unesco.org/images/0022/002277/227729E.pdf

United Arab Emirates University (2024) Home. https://www.uaeu.ac.ae/en/

University of Sharjah (n.d.) Facts and figures. https://www.sharjah.ac.ae/en/about/agc/Pages/Facts-Figures.aspx

Vallez M, Lopezosa C, Pedraza-Jiménez R (2022) A study of the web visibility of the SDGs and the 2030 Agenda on university websites. Int J Sustain High Educ 23(8):41–59

Waghid Y, Davids N (2014) On the (im)possibility of democratic citizenship education in the Arab and Muslim world. Stud Philos Educ 33:343–351

Yin RK (2003) Case study research: design and methods, 3rd edn. Sage Publications, New York

Zamora-Polo F, Sánchez-Martín J, Corrales-Serrano M, Espejo-Antúnez L (2019) What do university students know about sustainable development goals? A realistic approach to the reception of this UN program amongst the youth population. Sustainability 11(13):3533

Zayed University (2023) Fact book 2021–2022. https://allt-uae.zu.ac.ae/www-zu/open-data/wp-content/uploads/sites/2/2023/03/FactBook2021_22-i.pdf

Salem Alhammadi is currently pursuing a Master's in Education and Research at Geneva University. He holds a bachelor's in international studies from Zayed University. His academic interests inc Bachelor's degree in International Studies Bachelor's degree in International Studies lude international affairs and education, with a particular focus on the UAE.

Abdeljalil Akkari is a professor and the director of the research group on international education at the University of Geneva, Switzerland. He is also visiting professor at the Al-Farabi Kazakh National University, Kazakhstan. He is a regular consultant for UNESCO and other international organizations. He was the Dean of research at the Higher Pedagogical Institute HEP-BEJUNE (Bienne, Switzerland) and assistant professor at the University of Maryland Baltimore County (Baltimore, US).

Omani Universities' Engagement with the SDGs: Current Practices and Future Opportunities

12

Munira Khalifa Ali Al Wahaibi

Abstract

Higher education plays a crucial role in advancing Oman's progress toward the United Nations' Sustainable Development Goals (SDGs) and Oman Vision 2040, which emphasize a knowledge-based and sustainable future. Omani universities have officially committed to achieving the SDGs. However, coordination and mechanisms for monitoring progress and engaging stakeholders remain limited. This qualitative study examines how Omani universities are implementing SDG initiatives, focusing particularly on SDG 13 (Climate Action). It explores institutional practices and decision-makers' perspectives on enhancing effective, contextualized, and participatory approaches to SDG implementation. Data were collected through semi-structured interviews, event observations, and analysis of institutional documents. Findings highlight active collaboration among national committees representing higher education institutions and the Oman National Commission for Education, Culture and Science. The study underscores the essential role of higher education in advancing national SDG efforts and provides policy and practical recommendations to strengthen SDG integration. It concludes by encouraging regional replication of Oman's growing experience in aligning higher education with sustainable development.

Keywords

Higher education · Sustainable development · SDGs · Oman

12.1 Introduction

The Sultanate of Oman has demonstrated significant progress toward achieving the United Nations' 17 Sustainable Development Goals (SDGs). The government views these goals as a natural continuation of Omani societal values, including equality, peace, sustainability, good health and well-being, participation, and inclusion (National Committee for Achieving Sustainable Development Goals 2019). According to the *Sustainable Development Report 2022*, Oman ranked 81st out of 163 countries and first among Gulf Cooperation Council (GCC) states in overall SDG performance (Sachs et al. 2025). The *Arab Region SDG Index and Dashboard Report 2022* confirmed Oman's leading regional position, placing it ahead of the United Arab Emirates (85th), Qatar (94th), Saudi Arabia (96th), Kuwait (101st), and Bahrain (102nd). This progress reflects a consistent national commitment to sustainable development principles embedded in *Oman Vision 2040*.

M. K. A. Al Wahaibi (✉)
Sultan Qaboos University, Muscat, Oman
e-mail: munirak@squ.edu.om

K. Al'Abri and E. Buckner (eds.), *Higher Education and Sustainable Development in the Arab Gulf States*, Sustainable Development Goals Series, https://doi.org/10.1007/978-3-032-19559-3_12

Since the adoption of the SDGs in 2015, Oman has systematically integrated sustainability into its long-term national vision and planning framework (Oman Vision 2040 Implementation Unit 2025). The government established a National Committee for Sustainable Development (NCSD) to coordinate efforts across ministries and sectors and to develop a national monitoring and evaluation system to track SDG indicators. These initiatives are complemented by nationwide awareness programs—including workshops, seminars, and public consultations—designed to foster community engagement in sustainable development. Oman's 2019 Voluntary National Review (VNR) to the United Nations emphasized inclusive policy formulation through more than 40 consultative panels with youth, educators, and civil-society organizations, ensuring that Vision 2040 reflects the aspirations of Omani society.

Higher education institutions (HEIs) are central to achieving these goals. Universities and colleges serve as drivers of research, innovation, and social engagement that contribute directly to SDG 3 (Good Health and Well-Being), SDG 4 (Quality Education), and SDG 13 (Climate Action). Sultan Qaboos University (SQU), the nation's leading public university, has aligned its research agenda with sustainability priorities through initiatives in health sciences, renewable energy, and environmental management (Sultan Qaboos University 2023). In 2021, SQU organized an international virtual workshop in collaboration with UNESCO and the Oman National Commission for Education, Culture and Science to discuss strategies for integrating the SDGs into higher education curricula and institutional governance (Oman National Commission for Education, Culture and Science 2023). Speakers stressed that HEIs must play an active and measurable role in achieving the 2030 Agenda within the framework of Vision 2040.

Despite these positive initiatives, progress across Omani universities remains uneven. Al Abri and Al Hadi (2020) found that Sultan Qaboos University has been contributing to sustainable development mainly through its core functions of teaching, research, and community service but has not yet set explicit SDG targets or monitoring metrics. Comparable evidence from other Omani universities is scarce, indicating a knowledge gap between national policy intentions and institutional practice. Globally, studies indicate that this gap is common, as universities often endorse the SDGs conceptually, but they lack systematic implementation and reporting frameworks (Findler et al. 2019; Lozano et al. 2019). This situation highlights the need for a context-specific analysis of how Omani HEIs translate SDG commitments into action.

I observed this gap firsthand while integrating SDG-related content into a listening and speaking course for foundation-year students at SQU in 2021. Initially, none of the students were familiar with the SDGs; however, after a semester of active learning tasks and classroom discussions, students organized a public exhibition featuring artwork depicting local applications of the 17 goals. This experience illustrated the transformative power of education for sustainable development and the potential of Omani youth when equipped with knowledge and agency to contribute to national sustainability efforts.

This study holds significance in three key dimensions—national, academic, and social. At the national level, it aligns directly with Oman Vision 2040, which calls for a knowledge-based economy that fosters innovation, sustainability, and social cohesion (Oman Vision 2040 Implementation Unit 2025). By analyzing how higher education institutions engage with the SDGs, the study provides evidence-based insights for the Ministry of Higher Education, Research and Innovation and the Oman National Commission for Education to enhance policy coordination and resource allocation. It can also support the development of national guidelines for SDG integration across universities and colleges.

From an academic perspective, the study contributes to the limited body of research on higher education and sustainable development in the GCC. While global literature recognizes universities as pivotal actors in the 2030 Agenda (Findler et al. 2019; Lozano et al. 2019), empirical studies from the Arab region remain sparse. This research provides updated data on Omani universities' initiatives and challenges, filling a critical regional gap in scholarship. The results may also inform comparative studies across GCC countries, where efforts such as the UAE's University Sustainability Network and Saudi Arabia's Vision 2030 university programs illustrate institutional models that Oman could adapt (Emirates News Agency 2023; United Nations 2022).

Finally, the social and educational significance of this study lies in its advocacy for collective action. Education for sustainable development builds critical thinking, creativity, and ethical awareness (UNESCO 2025). Embedding SDG themes in Oman's higher education curricula can empower students to become agents of change. As climate change and social inequality are shared challenges, collaboration across universities and disciplines is imperative. Without broad engagement and awareness, Oman's progress toward the 2030 Agenda and Vision 2040 targets cannot be sustained.

12.2 Literature Review

The Sustainable Development Goals (SDGs) were launched by the United Nations in 2015. The SDGs encompass 17 goals and 169 targets, addressing a wide range of interconnected issues that are crucial for the overall development and well-being of a country in which the country can achieve long-term prosperity, equality, and a more sustainable future for their citizens. These goals aim at eradicating poverty and hunger; promoting good health and education; achieving gender equality; ensuring access to clean water and sustainable energy; building resilient cities; combating climate change; and fostering peace, justice, and strong institutions, among others, in pursuit of a sustainable and equitable future by the year 2030 (Transforming Our World: The 2030 Agenda for Sustainable Development|Department of Economic and Social Affairs n.d.).

Governments, organizations, and communities worldwide have drastically embraced these goals in their policies, visions, missions, action plans, and programs. In fact, the SDGs serve as a global framework for sustainable development efforts until 2030 (Resolution Adopted by the General Assembly on 25 September 2015. Transforming Our World: The 2030 Agenda for Sustainable Development n.d.).

12.2.1 Oman's Commitment Toward Integrating Sustainable Development Goals

The Sultanate of Oman, like other countries, has committed to honoring its international commitment to achieving SDGs by 2030. It considers the United Nations 2030 Agenda as a central pillar and as a framework to guide its national sustainable development efforts. In fact, Oman's Supreme Council for Planning (SCP) endorsed the SDGs in 2015 and since then, it has focused on incorporating and integrating them into its long-term development strategies, as outlined in the Oman Vision 2040, as well as into its short and medium-term action plans, strategies, and programs. It is worth noting that the Supreme Council for Planning (SCP) has been abolished as part of broader governmental restructuring aimed at aligning national institutions with the implementation of Oman Vision 2040, a long-term development plan designed to reduce dependence on oil revenues and diversify the economy. All the functions that are previously handled by the Supreme Council for Planning have been transferred to a newly re-established Ministry of Economy. The decision to establish a Ministry of Economy in place of the Supreme Council for Planning suggests a shift in focus toward centralizing economic planning and policy implementation within a dedicated ministry (Oman's Government Reshuffle -Trowers and Hamlins n.d.).

A dedicated organization called the Oman Vision 2040 Implementation Follow-up Unit was established to supervise the implementation and advancement of Oman Vision 2040. The unit is essential to ensure that the challenging goals of the vision are accomplished efficiently and on schedule. It is worth mentioning that the Oman Vision 2040 Implementation Follow-up Unit addresses global issues delineated by the United Nations' sustainable development agenda, while simultaneously advancing a sustainable and inclusive future for Oman by incorporating the SDGs into the framework of Oman Vision 2040 (Oman Vision 2040 Implementation Follow-up Unit n.d.). In relation to the SDGs, it is as follows:

Goal Alignment: Oman Vision 2040 covers a wide variety of development goals in the areas of the economy, society, and environment. Numerous of these goals are in line with the 17 Sustainable Development Goals (SDGs), which are designed to address issues like poverty, inequality, education, health, and climate change. The Oman Vision 2040 Implementation Follow-up Unit makes sure that Oman's national vision advances the accomplishment of these international objectives.

Integration of Sustainable Practices: The unit makes sure that the SDGs are incorporated into policies and programs created for Oman Vision 2040, with a focus on integrating sustainable practices into national planning.

Gathering and Tracking Data: The Oman Vision 2040 Implementation Follow-up Unit collects information and keeps an eye on key performance indicators to assess progress toward Oman Vision 2040 and the SDGs. SDG-aligned indicators may be used in this tracking process, which offers a framework for evaluating advancement toward both domestic and international development objectives.

Collaboration with Stakeholders: To guarantee a comprehensive approach to development, the unit works with a variety of stakeholders, including international organizations, partners in the commercial sector, and government agencies. To achieve the SDGs, partnerships are essential and this collaboration can help close gaps between local and global goals (SDG 17: Partnerships for the Goals).

An Omani national committee for sustainable development was formed in response to the SDGs and is chaired by the Deputy Chairman of the Ministry of Economy. It encompasses representatives from various development partners, such as government officials, academics, and members of the State and Shura Councils. It also includes representatives from the private sector and civil society whose participation is considered crucial at the national and local levels to ensure its ownership of the development agenda. The primary responsibility of this national committee is to oversee and follow up on the preparation of the Voluntary National Review (VNR), which is a report that provides an assessment of Oman's progress in implementing the SDGs. It provides information on how Oman is proceeding in the SDGs in which it points out its achievements, challenges, and future plans (Supreme Council for Planning 2019).

Alongside this committee, the National Center for Statistics and Information (NCSI) actively engages in monitoring and measuring indicators of the SDGs. The National Center for Statistics and Information (NCSI) is an important entity in the Omani government. It is responsible for collecting, analyzing, and publishing statistical data on various aspects of the Omani economy, society, and demographics. It provides accurate and reliable data to support evidence-based decision-making, policy formulation, and development planning in Oman (Chief Data Office - Omanuna Portal n.d.).

Regarding the SDGs, the National Center for Statistics and Information (NCSI) plays a vital role in compiling and managing data and indicators related to the SDGs in Oman. It ensures that all available data and indicators in Oman are gathered and incorporated into a comprehensive database. This database serves as a central repository for tracking progress on the SDGs. The center also develops a dedicated website specifically focused on the SDGs, which provides access to the collected data and information (SDG Site n.d.-a).

It can be deduced that the Omani National Committee for Sustainable Development and the National Center for Statistics and Information (NCSI) form a comprehensive framework that combines strategic oversight and data-driven monitoring of the SDGs in Oman. This framework contributes to effective decision-making, resource allocation, and progress evaluation in Oman's Sustainable Development Goals efforts. Yet, it is essential to recognize that neither can function in isolation from other Omani entities. Collaboration and synchronization are required across all levels, encompassing government and private sectors, stakeholders, and the Omani civil society. As a result, each of these entities has devised its own "institutional framework" to facilitate this harmonious cooperation.

12.2.2 The Role of the Institutional Framework

An institutional framework was established to supervise and manage the implementation and monitoring of the SDGs in each governmental entity. This involved designating a focal point within each institution to coordinate efforts, establish targets and indicators, and monitor progress toward achieving the SDGs.

The focal point acts as a central coordinating body responsible for managing the implementation of the SDGs across different sectors and ministries. It serves as the primary contact for both the National Committee and the National Center for Statistics and Information (NCSI). It is worth noting that the NCSI collaborates with the focal points that work in and communicate with relevant ministries and entities in order to ensure the availability of data on a regular and periodic basis. These focal points follow the methodologies provided by the United Nations to collect and report data on the SDG indicators (SDG Site n.d.-b).

Al Awati (2023) has pointed out that there are four assigned responsibilities for the focal points. They are coordination, establishing indicators and targets, monitoring and evaluating, and reporting.

Coordination First, in terms of coordination, the designated focal point coordinates efforts among various government ministries, departments, and agencies involved in SDG implementation. It facilitates collaboration, information-sharing, and the alignment of policies and actions to achieve the SDGs in order to ensure that there is a coherent and integrated approach to implementing the goals across different sectors.

Establishing Indicators and Targets Second, establishing indicators and targets specific to each SDG, which help track progress over time and identify areas that require additional attention and intervention, is the second responsibility. These indicators serve as benchmarks to measure progress toward achieving the SDGs. The focal point, in collaboration with relevant stakeholders, illustrates and defines targets that strike a balance between being ambitious and realistic and achievable.

Monitoring and Evaluation Third, the focal points establish a monitoring and evaluation system not only to track progress on the achievement of the SDGs, but also to identify gaps, challenges, and areas of success. Thus, they set up mechanisms for data collection, analysis, and reporting on the indicators established for each goal. Regular monitoring enables evidence-based decision-making and ensures that necessary adjustments can be made to policies and programs to achieve the desired outcomes.

Reporting Fourth, the focal point is responsible for preparing periodic reports on the progress in implementing the SDGs. These reports provide updates on the status of each goal, highlight achievements, and identify areas that require additional attention and investment. According to Al Awati (2023), the focal points share reports with the National Center for Statistics and Information to foster transparency, accountability, and knowledge sharing.

It can be understood and noticed that by establishing an institutional framework in each governmental entity, effective governance, coordination, and accountability in implementing the SDGs are ensured. This framework enables decision-makers and policymakers to align Oman's policies and programs with the global goals and track progress to achieve sustainable development.

It is noteworthy to mention that higher education institutions, similar to other governmental entities, actively work toward the attainment of not only SDG 4 but also the remaining 16 SDGs. The subsequent section will specifically highlight the role of higher education in contributing to the achievement of the SDGs.

12.2.3 The Role of Higher Education in Achieving the SDGs

Policymakers in Oman have placed a strong emphasis on higher education and have recognized its role in developing human resources and contributing to economic, social, and environmental development. In fact, the significance of higher education in Oman in driving progress toward the United Nations' 17 SDGs has been underscored, aligning it with the Oman Vision 2040. This alignment strives to lead transformative changes and build up a promising and prosperous future for Oman's knowledge-based economy and society.

As mentioned earlier, the fourth SDG focuses on ensuring inclusive and equitable quality education and promoting lifelong learning opportunities for all. It contains a total of seven targets which aim to address various aspects of education and human development. These targets endeavor to ensure that everyone has access to quality education regardless of their background or abilities. They also focus on creating safe learning environments, expanding educational opportunities through scholarships, and enhancing the availability and quality of teachers. By achieving these targets, stakeholders can work toward providing inclusive and equitable education for all (Role of Higher Education Institutions in Achieving SDGs Discussed 2021).

The universities and colleges in Oman offer various specializations and do contribute to sustainable development through research and community participation and integrate sustainable development concepts into their programs. Their educational systems promote tolerance, respect for others, and citizenship education, with curricula and textbooks designed to provide knowledge, skills, and values related to sustainability as they aim to raise a generation aware of sustainable development themes and values.

The Ministry of Education has developed programs and projects to instill values of citizenship and responsibility among children and young people.

Table 12.1 Universities and their contributions to the sustainable development goals (SDGs)

University/college	Focal SDGs
Sultan Qaboos university	SDG1, SDG2, SDG3, SDG4, SDG5, SDG6
Middle east college	SDG1, SDG2, SDG3, SDG4, SDG5, SDG6
A'Sharqiyah University	SDG1, SDG2, SDG3, SDG4, SDG5, SDG10
Sohar university	SDG4, SDG8, SDG9, SDG12, SDG16, SDG17
Dhofar university	SDG1, SDG3, SDG7, SDG10, SDG11, SDG12
Nizwa university	SDG1, SDG2, SDG5, SDG12, SDG13, SDG14
University of technology and applied sciences	SDG1, SDG7, SDG10, SDG11, SDG12, SDG13

Note SDG—sustainable development goal

Significantly, Sultan Qaboos University, Middle East College, A'Sharqiyah University, Sohar University, Dhofar University, Nizwa University, and the University of Technology and Applied Sciences are among the educational institutions that have been officially enrolled on the SDGs universities' platform. This digital platform, established under the auspices of the United Nations, acts as a platform for showcasing universities' efforts in advancing the 17 SDGs. Each of these seven universities has actively embraced and committed to promoting six specific SDGs (SDGS UNIVERSITIES n.d.) (Table 12.1).

Overall, Oman's educational policies and initiatives aim to develop a well-rounded and sustainable education system that meets the needs of its population and contributes to national and global development goals.

12.3 Research Objectives and Methods

The study aims to examine the existing practices, policies, and initiatives undertaken by the seven Omani universities to achieve the Sustainable Development Goals (SDGs), with a focus on SDG 13 (Climate Action). It also endeavors to identify and evaluate the future opportunities and strategic plans developed by decision-makers for effective and engaging implementation of SDGs, with an emphasis on customization to the local context in Oman, during the remaining 8 years to 2030. Specifically, this study answers the following research questions designed to bridge the knowledge gap between the strategic intentions and the actual practices of higher education institutions in achieving the SDGs:

1. What are the current practices that the seven Omani universities have adopted to achieve the SDGs in general and SDG 13 in particular?
2. What are the decision-makers' future opportunities and plans for effective, customized, and engaging implementation of the SDGs in the remaining 8 years of attaining the SDGs 2030?

To answer these questions, the study will investigate the specific strategies and actions implemented by universities to work toward the SDGs in general. It will highlight the practices adopted by these universities to contribute to SDG 13, which focuses on climate action. The findings may include various initiatives, programs, and projects undertaken by universities to address climate change and reduce their carbon footprint.

Identification of decision-makers' opportunities and plans: The study aims to explore the perspectives and intentions of decision-makers within universities regarding the future implementation of the SDGs. The findings may include the identification of opportunities and challenges faced by decision-makers in effectively implementing the SDGs within the given timeframe. It may also highlight the plans, strategies, and actions proposed by decision-makers to enhance the customized and engaging implementation of the SDGs in the remaining 8 years leading up to 2030. Based on the findings, the study may provide recommendations for universities and decision-makers on how to improve their practices and strategies for achieving the SDGs. These recommendations may include specific actions, policy changes, collaborations, and resource allocation suggestions to optimize the impact of universities' efforts toward the SDGs, particularly SDG 13.

To answer these questions, I adopted a qualitative method, specifically a descriptive qualitative design. Data were collected using two qualitative research tools: interviews and official studies and documents. The first interview was with Khalid Al Mawali who is an Assistant Director of the Department of Education Sector in Oman National Commission for Education, Culture and Science. The second interview was with Sawsan Al Awati who is the head of the statistical department at the National Center for Statistics and Information (NCSI).

I analyzed the strategic plans thematically to identify the lessons and insights that are gained from the current practices and future plans for achieving the SDGs in general and SDG 13 in particular.

12.4 Findings

12.4.1 A Partnership Between Universities and the Omani National Commission for Education, Culture and Science

Oman's National Commission for Education, Culture and Science is the counterpart to international, Arab, and regional organizations concerned with education, culture, and science (UNESCO, ISESCO, ALECSO) in the Sultanate of Oman (National Commission n.d.). It serves as a communication entity with these organizations to contribute to promoting their goals and encouraging the implementation of their programs within the Sultanate in coordination with relevant authorities. Additionally, it contributes to driving development through purposeful pioneering initiatives and disseminating information about developmental efforts in the Sultanate in the fields of education, culture, and science within the framework of the relationship with the three organizations. The education sector in this National Commission has three sections, namely, the School and Vocational Education Department, the Higher

Education, the Scientific Research Department, and the Statistical Data Department.

According to Khalid Al Mawali, the Assistant Director of the Department of Education Sector in the Oman National Commission for Education, Culture and Science, there is a national team responsible for monitoring and implementing SDG 4, which comprises seven distinct targets. Remarkably, Target 7 encompasses the comprehensive scope of all 17 SDGs. In other words, the educational institutions are bearing the responsibility not only for attaining SDG 4, specifically focused on education, but also for actively contributing to the accomplishment of the entire 17 SDGs. The team consists of members from the Ministry of Education, Ministry of Higher Education, Sultan Qaboos University, University of Technology and Applied Sciences, National Statistics Center, civil society, and private sector institutions such as the Environmental Association.

Al Mawali (2023) has asserted on the numerous efforts in organizing a series of workshops that target higher education institutions in the implementation and achievement of the SDGs. Among the workshops that are scheduled for October of this year is a workshop entitled "Education Laboratories for Sustainable Development Goals" and it is specifically targeting higher education institutions. The team is also working on initiatives that are related to green schools and campuses, digital transformation, and youth engagement in greening education. Al Mawali (2023) has pointed out that currently an initiative called "*Education for Sustainable Development*" has been introduced to the UN and he has illustrated that it is the only initiative of its kind in the Middle East.

12.4.2 University Initiatives

This section highlights the most valuable efforts a number of higher education institutions have performed in attaining the SDGs. They are Sultan Qaboos University, Dhofar University, Nizwa University and Muscat University.

12.4.2.1 Sultan Qaboos University

Sultan Qaboos University is the largest and oldest university in the country and is the country's flagship public university. It has undertaken deliberate efforts to align its institutional priorities with the Sustainable Development Goals (SDGs) and Oman Vision 2040, ensuring that sustainability principles are embedded across its strategic and operational frameworks. Its executive plan for the 2021–2025 period identifies five central pillars that structure this alignment. First, the university emphasizes the quality of its graduates and learning outcomes, seeking to equip students with high-level knowledge and skills that correspond to national and international labor market needs. Second, it prioritizes scientific research and innovation by fostering a culture of inquiry, supporting impactful projects, and encouraging both faculty and students to contribute to evidence-based solutions for societal challenges. Third, the university is committed to building partnerships with local and international institutions, industries, and communities to promote collaboration, knowledge exchange, and joint initiatives that enhance its societal impact. Fourth, it advances financial sustainability by diversifying revenue streams, optimizing resource management, and securing the university's long-term stability. Finally, the institution places strong emphasis on governance and institutional development, cultivating transparency, accountability, and continuous improvement across all levels of administration. As Al Abri (n.d.-b) notes, these five pillars encompass 20 projects, each with practical mechanisms and measurable performance indicators that together operationalize the university's contribution to the SDGs. Collectively, they illustrate how strategic planning and institutional governance can serve as key levers for integrating global sustainability commitments into higher education practice.

12.4.2.2 Dhofar University

Al Mashaikhy (n.d.) illustrates that Dhofar University has articulated six interrelated strands of work to advance the Sustainable Development Goals (SDGs) across its institutional mission. First, the university is intensifying research and scholarship on multiple dimensions of sustainability, including environmental protection, social inclusion, and economic growth. Second, it is actively cultivating creativity and innovation among faculty and students by providing mentorship, resources, and recognition for those developing novel ideas and practical solutions aligned with the SDGs. Third, Dhofar University is developing and implementing programs and projects designed to generate tangible impacts on sustainable development outcomes. Fourth, it is enhancing the curriculum by embedding sustainability principles across disciplines, ensuring that teaching and learning foster awareness and responsibility for sustainable futures. Fifth, the university is expanding open educational platforms to increase access to knowledge, raise awareness about the SDGs, and facilitate the exchange of ideas within and beyond the campus community. Finally, Dhofar University is linking education with community development by strengthening partnerships with local communities, government bodies, and non-profit organizations to collaboratively address sustainability challenges. Taken together, these six strands reflect a holistic and proactive institutional strategy that seeks to integrate sustainability into the university's structures and practices while empowering students to serve as catalysts for positive change in their societies and beyond.

12.4.2.3 Nizwa University

Nizwa University's commitment to the SDGs focuses on two areas, namely, its investment fund and its institutional strategic plan. Nizwa University has an investment fund that has played a crucial role in supporting the university's efforts to achieve the SDGs. As highlighted by Al Shukily (n.d.), the fund was established concurrently with the university itself. It comprises several existing companies, each contributing to specific aspects of SDGs.

One notable company within the fund focuses on promoting and developing green initiatives within the university, with a particular emphasis on utilizing renewable energy. This company aims to explore and implement sustainable energy solutions, contributing directly to SDG targets related to clean energy and environmental sustainability.

Another company within the fund is dedicated to natural landscapes and afforestation. This highlights the commitment of Nizwa University to preserving and restoring ecosystems, aligning with SDG goals concerning biodiversity conservation and sustainable land management.

Furthermore, Al Shukily (n.d.) has asserted that such companies have the potential to evolve into industrial ventures that leverage scientific research and innovation. For instance, the renewable energy company can harness research and technological advancements to create and implement sustainable energy solutions at a larger scale.

Along with the investment fund, the university has developed its strategic plan on a number of pillars. The following are the bases of the university's strategic plan, which endeavors to include the achievement of the SDGs. **Human resources** are the main pillar of the university. Therefore, training and orientation programs are offered to a large number of Omani youths in various fields in order to empower them to contribute to sustainable development. **Creating a conducive environment for research** that encourages faculty and students to engage in innovative and impactful research projects. **Developing policies, frameworks, and strategies** to govern and promote scientific research and innovation that are aligned with the principles of SDGs. **Establishing strong external collaborations** with external stakeholders, including government agencies, industries, and organizations in order to foster partnerships and knowledge exchange that support the SDGs.

12.4.2.4 Muscat University

Muscat University aims to contribute effectively to the educational landscape, national development, and the achievement of Oman's long-term vision. It has also built up its strategic plan on establishing international partnerships and collaborations, developing the student's character and skills and aligning it with the key pillars of Oman Vision 2040 and therefore achieving the SDGs (Al Mazrooghi n.d.).

12.5 Discussion and Recommendations

The current practices toward achieving the SDGs 2030 can be inferred from the data discussion. It can be inferred that there is a well-established and systematic coordination between the national teams within educational institutions and the National Commission at the national level. This will facilitate the process of gathering relevant information, conducting assessments, and help streamline the transferring of this information to the Ministry of Economy and ensure the timely completion of the second Voluntary National Review of the Sultanate of Oman 2023 which is currently in the process of being written.

One salient finding that should be emphasized is that the universities strategic and executive plans are well structured and grounded in education, research, and community engagement. These efforts contribute to achieving inclusive education, promoting lifelong learning, and advancing scientific research that supports the development of a knowledge-based society and competitive national talent.

Although the existence of challenges among these higher education institutions toward achieving SDGs is inevitable, the universities and colleges have found the suggested annual gathering for presenting their own mechanisms, best practices, and experiences in attaining the SDGs and Oman Vision 2040 to be constructive and efficient.

The series of workshops and initiatives demonstrates the significance of collaboration between higher education institutions and the Oman National Commission for Education, Culture and Science in contributing to sustainable development in Oman through the use of their respective expertise, resources, and networks. Such workshops and initiatives provide a platform for knowledge exchange, capacity building, and the development of practical solutions aligned with the Sustainable Development Goals (SDGs).

This study has exhibited the systematic and well-organized mechanism for ensuring the collaborative and transparent relationship between the Ministry of Higher Education and the Oman National Commission for Education, Culture and Science in their endeavors toward achieving the SDGs. In fact, they can effectively coordinate their initiatives, monitor progress, and ensure that actions taken align with the objectives of the SDGs and Oman Vision 2040.

This study has also endeavored to shed light on the current intentions, strategies, plans, and programs that Sultan Qaboos University, Nizwa University, Dhofar University, and Muscat University have set up toward achieving the SDGs. It has emphasized that higher education institutions play a pivotal role not only in achieving SDG 4, which focuses on quality education, but also endeavour to contribute to the broader set of Sustainable Development Goals (SDGs), with particular emphasis on SDG 13. These institutions play a multifaceted role that is essential for creating a sustainable future. This includes setting up various key strategies, intentions, and plans, which are educating and inspiring the next generation of leaders, conducting scientific research and innovation, fostering local and international partnerships, implementing sustainable practices, influencing policy, engaging and engrossing with communities, and building capacity and infrastructure.

It has concluded by providing some valuable insights and recommendations that will definitely pave the way for the higher education institutions in Oman over the remaining 8 years toward the successful achievement of the SDGs by 2030. Nonetheless, there are some limitations with this study. For example, I was not able to obtain information about the practices of three universities (Sohar University, A' Sharqiyah University, and The University of Technology and Applied Sciences) and one college (The Middle East College) that are registered in the SDGs Universities platform. It might be more beneficial to involve focal points within the higher education institutions. These focal points could be individuals responsible for overseeing specific programs, strategies, or initiatives related to the research topic. By engaging these focal points, researchers can gain deeper insights into the specific activities, approaches, and detailed information that may not be readily available through other sources.

Thereby enriching the overall understanding of the subject matter.

Although the higher education institutions have made significant efforts in implementing and achieving the SDGs in the Omani context, there remains a knowledge gap among teachers and students regarding the understanding of the 17 SDGs. In fact, the first Voluntary National Review in 2019 has pointed out that the educational institutions lack awareness about the SDGs (Supreme Council for Planning 2019). Consequently, it is crucial to disseminate information about these goals through appropriate communication channels to raise awareness and understanding.

It is not easy to change the higher education curriculum and programs to involve the SDGs. Yet, efforts can suggest a course as a university requirement in which these goals are embedded either explicitly or implicitly. In fact, Al Abri (n.d.-a) has exhibited that the role of higher education in sustainable development lies in transforming the means and processes of knowledge production and its research applications as well as the methods of student learning. Consequently, this empowers students and academics to be more socially responsible, critical, and sensitive to sustainability issues throughout their lives.

This is evident from this study, which involved implementing the Sustainable Development Goals (SDGs) in a listening and speaking course in 2021 in the Foundation Program at Sultan Qaboos University. At the beginning of the course, the participants were not familiar with the SDGs. After spending an entire semester learning about these goals, the students were able to conduct a creative exhibition in which they expressed their understanding of the SDGs through drawings that conveyed messages about how these goals can be achieved in Omani society.

The higher education institutions have the potential to create an ecosystem that fosters collaboration, knowledge-sharing, and collective action to advance the Sustainable Development Goals (SDGs) as a whole, with a specific focus on SDG 13. This ecosystem would facilitate the exchange of ideas, foster research collaborations, foster the development of sustainable solutions, propose the use of artificial intelligence (AI) in accomplishing the SDGs, and promote the sharing of best practices among stakeholders. They can contribute significantly to addressing the social, economic, and environmental challenges associated with the SDGs. By working together, higher education institutions can make a tangible impact in achieving SDG 13 and creating a more sustainable future.

References

Al Abri M, Al Hadi S (2020) Sultan Qaboos University and its contribution to the sustainable development goals in Oman. J Educ Sustain Develop 12(3):44–59

Al Abri K (n.d.-a) The development of higher education in the Sultanate of Oman and sustainable development [PowerPoint slides]

Al Abri K (n.d.-b) The strategic plan of Sultan Qaboos University 2016–2040 and its contribution to achieving the sustainable development goals of the sultanate [PowerPoint slides]

Al Mashaikhy K (n.d.) The role of higher education institutions in the Sultanate of Oman in achieving the sustainable development goals: Dhofar University [PowerPoint slides]

Al Mazrooghi Y (n.d.) The role of higher education institutions in the Sultanate of Oman in achieving the sustainable development goals: Muscat University [PowerPoint slides]

Al Shukily A (n.d.) The role of scientific research in sustainable development: Nizwa University [PowerPoint slides]

Chief Data Office – Omanuna Portal (n.d.) Chief data office. https://www.oman.om/wps/portal/index/gov/egovernment/cdo

Emirates News Agency (2023) Arab region SDG index and dashboards report 2023 unveiled. WAM—Emirates News Agency

Findler F, Schönherr N, Lozano R, Reider D, Martinuzzi A (2019) The impacts of higher education institutions on sustainable development: a review and conceptualization. Intern J Sustain High Edu 20(1):23–38. https://doi.org/10.1108/IJSHE-07-2017-0114

Lozano R, Barreiro-Gen M, Lozano FJ, Sammalisto K (2019) Teaching sustainability in European higher education institutions: assessing the connections between competences and pedagogical approaches. Sustainability 11(6):1602. https://doi.org/10.3390/su11061602

Ministry of Higher Education, Research and Innovation (2023) Higher education in Oman: Strategic initiatives and development plans. Government of Oman

Ministry of National Economy (2019) Voluntary national review: sultanate of Oman 2019. Government of

Oman. https://sustainabledevelopment.un.org/content/documents/23426VNR_2019_Oman_Report.pdf

National Commission (n.d.) Oman national commission for education, culture and science

National Committee for Achieving Sustainable Development Goals (2019) National report on Oman's progress toward the sustainable development goals. Government of Oman, Muscat

Oman National Commission for Education, Culture and Science (2023) Annual report on education for sustainable development. Ministry of Education, Muscat

Oman Vision 2040 Implementation Follow-up Unit (n.d.) About Oman vision 2040 implementation follow-up unit. https://www.oman2040.om/about/unit

Oman Vision 2040 Implementation Unit (2025) Oman vision 2040 strategy and implementation framework. Government of Oman, Muscat

Oman's Government Reshuffle – Trowers and Hamlins (2020) Oman's government reshuffle. https://www.trowers.com/insights/2020/august/omans-government-reshuffle

Sachs JD, Lafortune G, Fuller G, Iablonovski G (2025) Financing sustainable development to 2030 and mid-century: sustainable development report 2025. Sustainable Development Solutions Network, Paris

SDG Site (n.d.-a) Home. https://sdg.ncsi.gov.om/sdgwebsite/home

SDG Site (n.d.-b) Partnership for SDGs. https://sdg.ncsi.gov.om/sdgwebsite/PartnershipforSDGs

SDGS Universities (n.d.) Universities in Oman. https://sdgsuniversities.org/universities/search/?region=oman&sort=date

Sultan Qaboos University (2023) Contributions to the sustainable development goals (SDG 3, SDG 4, SDG 13). Sultan Qaboos University Press, Muscat

Supreme Council for Planning (2019) First voluntary national review of the Sultanate of Oman 2019. https://andp.unescwa.org/sites/default/files/2020-10/First%20Voluntary%20National%20Review%20of%20The%20Sultanate%20of%20Oman%202019.pdf

The role of higher education institutions in achieving SDGs was discussed (2021) Oman observer. https://www.omanobserver.om/article/3389/Local/role-of-higher-education-institutions-in-achieving-sdgs-discussed

Transforming our world: The 2030 Agenda for Sustainable Development (n.d.) United nations department of economic and social affairs. https://sdgs.un.org/2030agenda

United Nations (2022) The sustainable development goals report 2022. United Nations. https://unstats.un.org/sdgs/report/2022/

UNESCO Institute for Higher Education in Latin America and the Caribbean (IESALC) (2025) Higher education and sustainability: universities as catalysts for the SDGs. UNESCO, Paris

Wals AEJ, Peters MA (2022) Transformative learning for sustainability education in the Anthropocene. Educ Philos Theory 54(8):975–988

Munira Khalifa Ali Al Wahaibi, MA, is a Senior Language Instructor at Sultan Qaboos University, holds an M.A. in TESOL from the University of Southern Queensland. She has recently completed a Master's in Instructional and Learning Technology (ILT) with high distinction from the College of Education at Sultan Qaboos University. Her thesis, rooted in Design-Based Research (DBR), explores a Project-Based Learning (PBL) framework for integrating Artificial Intelligence (AI) tools into ESL academic writing. Her research interests focus on learner autonomy, motivation, learning styles, and the use of innovative technologies—particularly AI tools—to enhance ESL academic writing, foster project-based learning, and promote intercultural communication and global citizenship in alignment with the Sustainable Development Goals. She follows a student-centered philosophy and has presented at local and international conferences.

Urban Sustainability Education and the "Othering" of the City: Exploring SDG 11 in the GCC Context

13

Abdellatif Qamhaieh

Abstract

Although SDG 11sets out several fundamental principles to operationalize and measure the sustainability of cities, sustainability at the urban scale remains a somewhat elusive concept. This is particularly true in globalizing GCC cities, where sustainability is often perceived as a "technological solution" or an add-on. In this context, delivering meaningful ESD remains a formidable challenge, especially in educational programs focused on urban studies. Through the case study of course ARCH 423, delivered at the American University in Dubai, this paper explores alternative pedagogical approaches that can make the delivery of ESD more engaging and effective for students. The paper concludes that studio-based approaches, common in the design fields, could be expanded into other disciplines offering ESD, especially in the GCC region.

Keywords

Dubai · ESD · Gulf urbanism · SDG 11 · Sustainability · Urban studies

13.1 Introduction

Similar to the sustainability discourse in general, sustainability at the urban scale remains a rather difficult concept to quantify and understand (Li et al. 2018; Thomas et al. 2021). While the notion of sustainability in urban studies itself is not new, the adoption of the Sustainable Development Goals (SDGs) in 2015 provided countries with operational targets to better understand where they stand in their efforts to address the climate crisis. Out of all SDGs, SDG 11—Sustainable Cities and Communities, is particularly important for cities since they are a significant contributor to carbon emissions (UN-Habitat 2020). Effectively, SDG 11 provides governments, city officials, researchers, and relevant stakeholders with a blueprint for improving cities and reducing their carbon footprint. Though most governments around the world adopted SDGs and set long-term plans to achieve these targets, measuring success and translating the principles into an "observable" reality remains rather complex, costly, slow, and sometimes lacks political will (Huang et al. 2015; Klopp and Petretta 2017; Thomas et al. 2021).

As the scope of the climate crisis continues to be alarming and is expected to worsen, the need for the education sector to respond has become even more critical. In 2019, UNESCO introduced the global framework for Education

A. Qamhaieh (✉)
American University in Dubai/United Arab Emirates, Dubai, UAE
e-mail: aqamhaieh@aud.edu

K. Al'Abri and E. Buckner (eds.), *Higher Education and Sustainable Development in the Arab Gulf States*, Sustainable Development Goals Series, https://doi.org/10.1007/978-3-032-19559-3_13

for Sustainable Development (ESD) (UNESCO 2020)—building on earlier efforts to introduce sustainability education at all levels, especially in Higher Education (HE) institutions. Most governments around the world have embraced this framework, and educational reform has been a strategic priority- including in the GCC region, albeit with varying degrees of success (QAA and Advance HE 2021; Zaidan et al. 2019).

Against this backdrop, this paper reports on the outcomes of the course *ARCH 423—Sustainable Urbanism*, which is part of the architecture program curriculum at the American University in Dubai. As a course built around SDG 11, it adopts hands-on design and art-based teaching methodologies to deliver its content. The paper explores whether such methodologies, common in the design fields, can help further ESD education, especially in the region. It begins by providing a theoretical background relative to ESD education in urban studies and architecture, and discusses sustainability at the urban scale within GCC cities. Methodologically, the paper reflects on the results of course assessments represented by the students' creative work, and where sustainability is highlighted through "*their own eyes*." The paper concludes that studio-inspired teaching methodologies could effectively deliver ESD education—even outside of the design disciplines.

13.2 Teaching Sustainability in Architecture and Urban Studies

Within well-established architecture programs globally, the discussion and educational modules focused on sustainability have been at the core of architectural education since the 1970s (Wright 2003). Historically, architecture as a profession used to be seen as "one with nature," while vernacular buildings were celebrated as an embodiment of sustainability. It was not until the *Modern Movement* in the twentieth century that architects pursued technology and industrialization ideals, glorifying dominance over nature and the motto of "form follows function." With that, the architectural profession and education departed from their ecological roots, only to be re-questioned again as shortcomings of modernism in design and city planning became more apparent (Deviren and Tabb 2014).

Recently, the debate about sustainability in architectural programs has resurfaced due to the ongoing environmental crises, and the subsequent rigorous sustainability requirements introduced by governments and professional accrediting bodies globally (Al-Naqbi and Alshannag 2018; Deviren and Tabb 2014). For example, in the US context, the National Architectural Accrediting Board (NAAB) mandates that programs instill environmental ethics in their graduates. The board also requires that programs and students demonstrate these learning outcomes as a requirement for granting accreditation (NAAB 2020). Similarly, in the UK, the Royal Institute of British Architects (RIBA) references environmental issues in four of its six core values. Therefore, architectural programs must demonstrate such commitment as a condition for licensure (RIBA Education 2021). Most local accrediting bodies in the GCC region also include references to sustainability in their accreditation and licensure checklists as part of their countries' commitments to SDGs and ESD (Al-Naqbi and Alshannag 2018)—though how these requirements manifest in GCC-based programs is a different question altogether (Benkari 2013).

Overall, the approach to introducing sustainable education in architectural programs varies from one program to another. While some offer separate lecture-based sustainability courses, others consider sustainability a natural ingredient of the design studio. Nevertheless, most architectural programs define sustainability through the narrow *building performance* lens and invest in teaching environmental simulation software to address this issue (Boarin et al. 2020). Recently, the requirements of professional fields appear to have had the most impact on architecture program curricula. The demand for sustainability and green-building professionals in Architectural firms has grown steadily in the last decade. Whether for sustainability concerns, bragging rights, or political gains, most governments, including those in the GCC, have introduced green-building certification systems such as LEED and BREEAM into their building regulations. This has forced educational programs to respond and adapt to keep their graduates professionally relevant and employable (Altomonte 2009).

When shifting gears toward substantiality in urban studies and at the urban scale, the picture becomes far more blurry. This "blurriness" refers to both sustainability education and the definition of sub-disciplines within urban studies as independent areas of knowledge (Bina et al. 2016). For starters, urban studies is an umbrella term that may include a focus on the built environment (represented by urban design, landscape, and architecture), but it may also include disciplines focusing on social studies, such as urban geography, ecology, and economics (Bina et al. 2016). Due to this broad definition, the focus of urban studies programs varies by context. It is also highly influenced by external factors related to the overall "ecosystem"—i.e., governance systems and their subsequent institutions (democratic vs. autocratic system), local urban problems (slums, flooding), historical issues (race, migration), and other elements that help shape the focus of the discipline (Levy 2015). In the American context, for example, academic programs in urban studies tend to focus on policy and socioeconomic issues, which naturally gives them a broader topic coverage. Meanwhile, in the European context and much of the Global South, urbanism programs are more preoccupied with design aspects. They are therefore often closely related to architectural programs, with a narrower focus on design—albeit with some exceptions.

Overall, regardless of program type, the delivery of sustainable education in urbanism appears inconsistent, follows diverse pedagogical pathways, and achieves varying degrees of success (Maruna 2019; Moosavi and Bush 2021; Park et al. 2022; Sinakou et al. 2022). Park et al. (2022) argue that while the collaborative nature of urban studies and the action/studio-based learning model lend themselves well to ESD education, these advantages are undermined by instructors' lack of understanding of ESD or a lack of commitment. Similarly, Bina et al. (2016), in their study of international graduate programs in urbanism, found that commitment toward ESD and the engagement with its complexity remain weak. Examining the local focus of urbanism programs, Hurlimann et al. (2021) analyzed several cases within the Australian context and found that ESD instruction tended to emphasize issues of social justice and social policy, while paying less attention to the broader climate debates.

Li et al. (2018) argue that most urbanism-related programs regard the notion of urban sustainability as vague and elusive, which limits its ability to inspire practical action. The authors also highlight the importance of the local context in shaping how academic programs engage with sustainability. In his seminal work, Jabareen (2012) developed a framework for studying sustainability within urban studies. He argued that there was considerable confusion between environmental education and sustainability education. Also, the three E's of sustainability—environment, economy, and equity—are too vague and poorly connected to sustainability at the urban scale. The author proposes an alternative framework for studying sustainability in urban studies that includes ethical, social, economic, ecological, and political dimensions, alongside spatial design aspects (Jabareen 2012).

Pedagogically, a broad consensus emerges in architecture and urbanism relative to the ideal setting for ESD delivery: The design studio environment (Bina et al. 2016; Jabareen 2012; Maruna 2019; Moosavi and Bush 2021; Park et al. 2022; Sinakou et al. 2022). The studio has long been at the core of design education and art due to its collaborative, iterative, and hands-on learning nature (Green and Bonollo 2003). Historically, disciplines outside the design field, especially engineering, viewed the studio environment as lacking academic rigor (Green and Bonollo 2003; Schön 1984). Therefore, these disciplines bypassed studio learning in favor of classical lecture-based courses. Despite recognizing the importance of studio-based pedagogy, a comprehensive study of international urbanism programs by Park et al. (2022) found that while programs employed multiple teaching methodologies, they still relied predominantly on the classic lecture to deliver ESD. The authors further argue that the additional effort and training needed to deliver ESD effectively contribute to the persistent challenges of embedding it within urbanism education.

Bina et al. (2016) reflected similar sentiments and argued that the inherent complexity and difficulty in defining sustainable urbanism make the delivery of ESD challenging. However, the authors still suggest that the multidisciplinary and integrative nature of urban studies should better align the discipline with ESD. Jabareen (2011, 2012) suggests that sustainability education should be taught in theory and practice due to its complexity. The author highlights the importance of teamwork and individual action research in helping students understand sustainability's multi-faceted nature. Jabareen also argues that sustainability education, in general, should take the form of an "*active engagement, and not a mere study of abstractions*" (Jabareen 2012, Line 6). Moosavi and Bush (2021) discuss that incorporating ESD in an interdisciplinary studio helped students grasp the principles more effectively. Applying urban sustainability concepts to a complex site encouraged creativity, shared ownership of the work, and negotiation, and fostered experiential learning among students. The authors also argued for collaborative learning, real-world scenarios, games, teamwork, and reflection through participation as essential ingredients for the success of ESD. Although other project-based and collaborative approaches are used, the authors contend that restrictions in the materials used to deliver ESD and a lack of comprehensive understanding of sustainability limited the success of the projects in their case study (Moosavi and Bush 2021).

13.3 The Urban Sustainability Discourse in the GCC

According to the UN-Habitat (2020), countries of the GCC are some of the most urbanized areas in the world, with more than 80% of the population living in urban areas. This pattern of urban growth is expected to continue as countries and cities experience rapid expansion driven by heavy investments in tourism and knowledge-based economies. These changes occur as GCC countries seek to diversify their economies beyond oil revenues (Zaidan et al. 2019). Urban development, especially in extreme climatic areas, comes at a high environmental cost. For example, the UAE and Qatar have some of the highest ecological footprints in the world (Crot 2013; Rizzo 2017). Large carbon footprints, high energy subsidies, low prioritization of environmental sustainability, and complex socio-political realities profoundly impact the region's sustainability debates (Zaidan et al. 2019). Additionally, the competitive nature of GCC countries, mainly as they strive to attract global capital, has limited cross-border collaboration, which is essential for combating sustainability issues (Al-Saidi 2021). While the GCC region performs poorly on sustainability indicators globally, it is in a better position than most Arab countries. Their engagement with international agencies and global actors has further propelled the GCC countries' sustainability agenda (Al-Saidi 2021).

In academic circles, most urban scholars examining sustainable development within the region's cities reflect a negative outlook and argue a mismatch between sustainability rhetoric and reality. For example, Rizzo (2017) argues that sustainability-focused projects in Qatar are branding exercises intended to put the country on the world map, while the true meaning of sustainability and the inclusion of all aspects of society are largely ignored. Similarly, Crot (2013) examined Masdar City in Abu Dhabi and argued that some tenets of sustainability—mainly economic, and to a lesser extent environmental—took precedence over social issues. Glitzy and glamorous "sustainable" projects, the author contends, represent an overreaction to unsustainable lifestyles and petro-dollar spending. According to Crot (2013), these projects represent *greenwashing* or an attempt to improve the country's image in response to international pressure and engagement (Crot 2013).

Rizzo and Petruccioli (2023) argue that most regional urban sustainability discussions are demonstrations of government power. They tend to reflect technology-based approaches to sustainable development, while other aspects of sustainability are often overlooked. In her work examining Masdar City in Abu Dhabi, Günel (2020) describes the resulting city as a "spaceship in the desert" and says that Masdar and similar projects represent a technical solution to address sustainability concerns but avoid dealing with complex social and economic issues. Gunel argues that these spaceship projects are devoid of political and cultural constraints and represent an imaginary notion without connection to actual conditions in the broader society and region (Günel 2020).

Reflecting similar themes, Al-Saidi and Zaidan (2020) and Zaidan and Abulibdeh (2021) discuss how the concept of master planning, or "master plans," has been used repeatedly in the GCC region to showcase innovation, technology, and alternative approaches to urban sustainability. Smart, green, carbon–neutral, and sustainable cities are common terminologies in the region. These terms often refer to projects varying in size and scale with a "thematic" concept behind them, highlighting sustainability. Master planning is common in the GCC region as a form of urban development and is often considered the hallmark of modern Gulf urbanism. These mega-projects are also essential city branding strategies and are seen as attempts to diversify economically. They also act as vehicles for attracting tourism, investment, and global talent (Elsheshtawy 2009; Zaidan and Abulibdeh 2021).

Meanwhile, Al-Saidi (2022) examines the SDG indicators within the GCC region and their effectiveness in reflecting real progress. He argues that the GCC countries face similar critical challenges related to urban development, including water scarcity, large ecological footprints, and ecosystem destruction. The author suggests revising the SDG indicators in favor of customized indicators for the region, as the current reporting mechanism could indicate false progress in some areas. Al-Saidi also states that the noted progress on the SDG agenda has little impact or meaning on the daily, consumerism-focused lifestyles, and urban development patterns. He argues for a more refined "mini agenda" to be developed that focuses on issues such as equity and rights and prioritizes environmental concerns. The author also highlights the weak focus on ESD in the region as a major priority for policymakers (Al-Saidi 2022).

Within the UAE, the conditions of sustainable development share challenges similar to those of the broader GCC. Dubai, in particular, has spearheaded large-scale urban development in the region and became "instantly" the model everyone emulated (Bagaeen 2007; Ponzini 2011). It hosted Expo 2020, focusing on sustainability, and recently hosted the COP28 conference in the same site (renamed Expo City). According to the official reports, the UAE is on the path to achieving its UAE 2050 Energy Strategy, The UAE 2050 Climate Change Plan, and the 2030 Agenda for Sustainable Development (MoCCAE 2017; National Committee on Sustainable Development Goals 2018). Government agency reports indicate significant progress toward meeting international commitments and sustainability goals (Barhouma 2023). However, limited academic research on the UAE's progress in its urban sustainability initiatives was found while preparing this manuscript.

13.4 Urban Form in Dubai

Dubai is an important city in the region that has risen to prominence rapidly due to its considerable investments in infrastructure, signature architectural projects, city branding exercises, and economic liberalization (Elsheshtawy 2009; Molotch and Ponzini 2019; Pacione 2005; Ponzini 2011; Ponzini and Alawadi 2022). The story of Dubai's rapid urbanization has become famous, evolving from a small fishing town in the early twentieth century into the global metropolis it is today (Elsheshtawy 2009). Beyond this, Dubai continues to impress and successfully attract the world's corporations and economic elites, seeking to establish a foothold in the region. The Dubai urban development model has become somewhat of a template for surrounding cities, with other regional players such as Doha, Abu Dhabi, and Riyad learning lessons from the emirate's success

(Elsheshtawy 2008a; Ponzini 2011; Rizzo 2014, 2019).

Historically, the urban form in Dubai went through different stages of development (Alawadi 2017; Alawadi et al. 2018). From the small settlement next to Dubai Creek to the sprawling city of today, several policies and factors have shaped the city—including adopting neoliberal economic policies and creating free-economic zones (Elsheshtawy 2009). The term "Bigness" has been used to describe the latest trends in development, where glamor, spectacle, and large-scale projects reflect the mood in Dubai today (Alawadi 2017). For the most part, urban development is spearheaded by large private/semi-governmental influential development firms, especially outside of the city's older urban areas, and the city has no official central urban planning agency. In reality, multiple agencies regulate different parts of the city; meanwhile, large developers own large swaths of land, effectively setting their own planning agendas and regulations. This rapid form of development comes with a tremendous environmental impact due to the extreme financial cost, the consumption of resources and land on a grander scale, and the infringement on coastal ecosystems (Zaidan et al. 2019).

Similar to the development patterns in the wider GCC, most residential neighborhoods and government housing projects are modeled around the North American suburban model, with car-centric sprawling developments. The results are an ever-increasing automobile dependence, low-density suburban development patterns, and significant community segregation (Elsheshtawy 2008b; Khalaf 2006; Qamhaieh and Chakravarty 2020). Dubai has the usual socio-demographic challenges widespread in the Gulf, with many migratory workers living in various peripheral areas and primarily absent from the city's urban planning debate (Ewers and Dicce 2016). These delicate demographics and questions of *"right to the city"* complicate the sustainability discourse even further. The demographic divide creates somewhat parallel experiences with contrasting narratives of the city depending on social class. The important work by Yasser Elsheshtawy, who examines informality in Dubai, vividly captures such alternative narratives and realities (Elsheshtawy 2008c, 2010, 2013).

Despite Dubai's development and demographic challenges, the city is moving slowly but surely toward a more sustainable existence. The introduction of the Metro system in 2009, followed by the tram system, and the introduction of more sustainability-focused planning regulations are meaningfully and organically advancing the sustainability agenda, especially as the city becomes more important globally. Also, lessons from COVID-19 helped bring issues of urban health and quality of life to the forefront. Authorities recently introduced the Dubai 2040 master plan to make the city more livable and greener (RTA 2021). While the subtle but meaningful change-of-course results are yet to be seen, media outlets, newspapers, government press releases, and international design consultancies still push glitzy, cost-prohibitive, technology-focused, "spaceship" like takes on sustainability. Though these narratives are designed to feed into the city branding efforts and primarily for the world to see, this paper argues that they contribute to a negative, if not a vain, image of Dubai and detract from genuine progress on the urban sustainability front.

13.5 Research Methods

This paper adopts a qualitative instrumental case-study methodology (Kekeya 2021; Yin 2018). It reports on the student assessment outcomes for the course ARCH 423—sustainable urbanism, an elective course in the Bachelor of Architecture program at the American University in Dubai, typically delivered annually. The case-study approach was selected because the assessment methods were primarily exploratory, aiming to reflect on methods suitable for engaging students with sustainability at the urban scale and delivering more effective ESD methodologies. ARCH 423 was originally a lecture-based course; it has been purposefully restructured into an active learning environment (Prince 2004)—to advance its learning outcomes and increase students' engagement with the topic.

The sample results from students' assessments and the instructor's reflections were collected/noted over two consecutive course offerings. The participating students (a total of 50 students, spanning two academic years) were mainly senior architecture students, typically from Arab and Emirati backgrounds. As students in a private educational institution in Dubai, most came from affluent socio-economic backgrounds. The course was redesigned to include several readings, discussion sessions, significant fieldwork periods, artistic works, and project-based design exercises centered around SDG 11 principles. It was also delivered in a hybrid seminar/studio format to enhance the learning experience and align with key ESD competencies (QAA and Advance HE 2021).

The course assessment methods were used primarily to reflect on the students' grasp of SDG 11 principles. They were intended to increase in complexity, scale, and involvement as the academic semester progressed, and understanding of concepts increased. Due to the subjective nature of design and artistic works, grading rubrics were employed to evaluate student attainment. The rubrics, while varying from assessment to assessment, typically evaluated students' depth of contextual understanding, analysis of various sustainability themes, sensitivity to human-centric approaches, and creative (artistic) expression and critical thinking. The main assessments included:

Photo Essay Assessment: This task was focused on experiential learning. Students were required to participate in a journey through the city, using public transportation. They visited older areas within Dubai, explored different spaces, and interacted with users of these areas. The students were provided with a list of keywords that represented thematic areas of sustainable urbanism. They were then expected to capture images representing these themes. As a final step, the students were supposed to digitally (and artistically) manipulate their photos to highlight the keyword(s) intended by the images.

Reflective Architectural Collage Assessment: The students were tasked with reproducing key urban sustainability themes through architectural collage and artistic graphical representations. Architectural collage is a technique that was adopted by some of the modernist architects, where images or image fragments are joined together to generate new meaning and inspire imaginative thinking (Shields 2023). For the students, this assignment typically took place in the later part of the academic semester, when they had a better understanding of the course themes. The topics represented in this assignment revolved around areas of sustainable transport, access to urban green spaces, planning equity, housing, and other various areas of urban sustainability.

Project-based Assessment: This was the main design assignment in the course. It was also where all the different sustainability concepts were to be integrated into a final design proposal. The students worked in teams, and the projects involved significant fieldwork within pre-selected, predominantly lower-income and often neglected areas of the city. Students were tasked with proposing sustainable design solutions responding to the neighborhood conditions and sensitive demographic compositions. These designs were expected to be theoretically grounded in SDG 11 theory and to respond to course discussions. The completed designs were to be presented in the form of a neighborhood retrofit and upgrade proposal, following inclusive and human-centric design principles. Such participatory and context-sensitive methods were a departure from the typical top-down, high-end focus commonly seen in the city.

The work of the students was routinely evaluated through peer evaluations, external jury members, and by the course instructor. The best representative samples were collected and shared at the department level and on social media platforms. Students were asked to provide feedback on assignments after each submission (informally), and provided course overall feedback at the end of the academic year to validate results. Both external jury members and student evaluations indicated that they appreciated the assessments used in this course and the learning approach.

13.6 Findings and Discussion

In this section, an overview of student work samples is provided to demonstrate the assessment methodologies and the students' understanding of sustainable urbanism and SDG 11. Selected samples from the assessments are discussed below.

Sample Results from Photo Essay: An integral part of this assignment is the experience of visiting older, often run-down areas of the city. The course instructor's selection of these areas is purposeful and directly related to highlighting SDG 11components and ultimately contributing to ESD. This assignment highlights two important points. First, despite their run-down appearance, older parts of the city and neighborhoods hold inherent value. Second, the lower-income groups living in these areas (primarily migratory workers) are of equal importance and deserve the attention of city planning officials. Considering the context of Dubai, where luxury and glamor are celebrated and where the official discourse on sustainability involves lavish and often high-tech solutions, these are valuable lessons about "bottom-up" sustainability at the city scale that focus on the usually overlooked social dimensions of SDG 11. Most students participating in the course typically have either never visited these areas or tend to avoid them altogether, as they are perceived to cater to different *"other"* demographics.

Furthermore, it is mandated that the students travel to study areas using public transportation, such as the nearby metro. While central to SDG 11 and sustainable urbanism, using public transportation is, for most students, a new and often exotic experience. For some, it is even considered a taboo, evoking negative imagery associated with gender sensitivities and social class tensions. It is common for certain students to resist such a requirement, suggesting instead that they drive directly to the sites—or even alongside the metro. Interestingly, it is usually those who have never used public transportation who make these requests, a trend also noted in transportation planning literature, where non-riders often hold the most pronounced prejudices against public transportation (Ramos et al. 2019).

Encouraging public transport usage remains a consistent theme throughout ARCH 423 due to its environmental benefits, particularly in the automobile-dependent society in Dubai and the wider GCC. For the photo assignment, Figs. 13.1, 13.2, 13.3, and 13.4 showcase selected images generated by the students and later manipulated digitally to highlight their understanding of urban themes. In the author's view, this technique bridges theory and practice, allowing students to notice urban elements they typically overlook. Experiencing these elements in person and within context also deepens the students' understanding of their value. Individual commentary on the assignments is provided in the image captions below.

Results from the Collage Assessment: In these assignments, students reflect artistically on course themes, aiming for deeper meaning beyond what initially meets the eye. These collages/photomontages are required later during the semester and reflect the overall understanding and engagement with SDG 11-related issues. Students have complete freedom in choosing the focus of their collages, and the resulting work usually reflects topics that resonate personally with them. Some of the best collages often address sensitive issues such as diversity and rights within the context of the Gulf city. Figures 13.5, 13.6, 13.7, and 13.8 represent sample collages developed by the students, covering various themes of urban sustainability.

Results from the Project-Based Assessment: The most critical and effective course assessment is usually the design project, which serves as the final assignment where most learning is expected and demonstrated. Students build on extensive experiential learning, including site visits, on-site discussions, and interaction with area residents through informal conversations and surveys when possible. As discussed earlier, these project-based exercises provide excellent opportunities for problem-solving, critical thinking, and reflective thinking—essential components of successful ESD.

Fig. 13.1 Photo essay sample: the student highlights pedestrian-friendly urban infrastructure and sustainable architectural features in alignment with the tenets of SDG 11

Fig. 13.2 Photo essay sample. The students highlight biking as an active mobility solution and colorful shopfronts that reflect urban vitality, aligning with SDG 11

Fig. 13.3 Photo essay sample. The students highlight urban infrastructure and shading elements that encourage walkability and street life

Fig. 13.4 Photo essay sample. This image highlights shared bedspaces/rooms, common in Dubai's lower-income areas. Shortages in affordable housing are widespread in GCC cities and are a significant challenge to planning equity and social sustainability, yet they receive little attention in formal sustainability debates

Fig. 13.5 Collage assignment sample. The image highlights various SDG 11 and climate change-related issues, including sustainable transport, industrial pollution, housing problems, and active mobility

Most instructor-selected project sites purposefully involve older neighborhoods in Dubai, typically home to lower-income groups and informal housing arrangements. While some students initially might feel uneasy about working in these locations, the more they engage with the context, the better they grasp the interplay of the various components of urban sustainability, such as mobility and transport challenges, economics, infrastructure, loss of historical heritage, and planning equity (or the lack thereof). Rather than reading about these issues or hearing about them from the instructor, students witness them first-hand and respond by proposing design solutions or urban interventions. These projects close the loop in ESD by translating knowledge into outcomes and action.

Figure 13.9 below represents a project developed by the students for the *AL-Karama* neighborhood in Dubai, an ethnic enclave of historical significance. The neighborhood was part of an international student competition focused on revitalizing older neighborhoods in several cities globally. Students spent time in the field and

Fig. 13.6 Collage assignment sample. This submission highlights local heritage, urban farming, and e-mobility, all closely related to SDG 11

created a simple, human-centered design proposal based on field observations and a broad understanding of SDG 11. The submission stood out in its simplicity and focus on local community needs, contrasting with typical top-down interventions and other competition entries. The proposal was well received by the competition jury, resulting in a winning entry that was subsequently featured in official Dubai government media. The student proposal demonstrated a balanced approach, combining creativity with context sensitivity, as described below.

13.7 Conclusions and Implications

This chapter presents a brief overview of ESD education in architecture and urbanism using the results of course ARCH 423 as a case study. The paper examined the state of ESD within these disciplines. It then explored the sustainability discourse within the GCC and provided a snapshot of urbanism in Dubai. The paper highlighted several assessment methodologies and innovative approaches that appeared to enhance students'

Fig. 13.7 Collage assignment sample. This collage highlights diverse demographic groups within Dubai, featuring various city backdrops and experiences. It particularly spotlights the contrast between wealthier residents and lower-income worker groups, represented by worker housing imagery on the right-hand side

interest and engagement with urban sustainability issues—an area particularly challenging for them to relate to and understand, especially outside of specialized graduate studies.

Urban studies, as an umbrella term, naturally lends itself to discussions about sustainability due to the discipline's comprehensive nature, particularly when considering urban planning with its multiple policy and design aspects. It is also a relevant focus for ESD, considering the profound impact of urban development on the natural environment, as highlighted by SDG 11, and the significant amounts of energy and infrastructure needed to sustain the world's cities. However, even within urban studies, the various types of academic programs and their multiple areas of focus, ranging from design to policy, make this a nonlinear process, and much remains to be done to boost ESD within this field.

Fig. 13.8 Collage assignment sample. In response to Dubai's heavy automobile dependence, primarily along the city's central highway (Shk. Zayed Road), this collage artistically highlights a more human-centric, imagined version of the city

Within the GCC region, most urban studies-related programs approach sustainability primarily through the lens of architecture and design, and very few programs have a more comprehensive urban policy outlook outside of some specialized graduate programs. This issue is complicated by the job market, where students with design backgrounds are favored, and job opportunities are much more flexible. Furthermore, educational programs in design often lack faculty with an in-depth understanding of sustainability, particularly at the urban scale. Most are unfamiliar with ESD, so capacity building and training among faculty members could be another significant area of concern.

In Dubai, the fragmented and car-centric urban patterns of development make discussions about urban sustainability appear abstract and remote, especially for younger students. More importantly, how can such students relate to sustainable, human-centered environments when they have never directly experienced them? On the one hand, the official narrative suggests, for example, that a glamorous 93 km air-conditioned, glass-covered running and cycling track is a model of sustainability (Skirka 2023), and that technology-heavy, expensive proposals such as Masdar are ideal templates to emulate. On the other hand, these younger generations have been conditioned "organically" by their peers and society to live on the periphery as casual observers of the city: their

Fig. 13.9 Images from the winning project: **Left frame**—The image proposes improvements to a run-down location within Al-Karama, mainly used as housing for low-income workers. While most formal interventions in the neighborhood call for demolition, students, based on fieldwork and course discussions, understood the housing needs of these lower-income segments and opted to upgrade the area while maintaining existing tenants. **Upper right frame**—The image highlights human-centric design. The students' proposal includes developing an outdoor cinema and park, reflecting their discussions with residents, who felt somewhat neglected with nothing to do after a long day at work. Most of these older areas are often ignored in planning efforts, which is why the emphasis is on equity. **Lower right frame**—Attention to detail in color and material helps boost the neighborhood's image and overall quality. Students also address mobility concerns by providing clearly marked bath routes and pedestrian-friendly interventions

daily routines gravitate around cars (Qamhaieh and Chakravarty 2020), shopping malls, coffee shops, and movie theatres. They also have been "told" that the outdoors is hostile, too hot, or culturally inappropriate for walking. Moreover, public space and public transport are perceived as not for them, but for the others. For these students, the city could be reduced to a few sparse destinations, visited mainly by car, with little value in between. With this in mind, the discussions about urban sustainability, or even the SDGs, come across as academic, theoretical, and detached. To complicate matters further, the delicate demographic composition within GCC cities—with visible competition between nationals and expatriates, alongside the precarious living conditions of large numbers of low-income migratory workers—contributes to concerns of social equity and representation (Ewers and Dicce 2016, 2018; Rizzo and Petruccioli 2023). Residents of these cities, including faculty in local universities, often feel uneasy or cautious about addressing such issues, which means that discussions of sustainability remain, at best, partial.

In conclusion, making sustainability understandable and relatable to students remains challenging in any context, but even more so in the GCC. The example of ARCH 423, as demonstrated here, could have significant and valuable lessons relevant to ESD and the sustainability discourse more broadly. Given the complex and contested nature of sustainability and the multiple disciplines that contribute to its definition, traditional, lecture-based pedagogical approaches are often unsuitable for ESD, as they render its concepts abstract and challenging. Meanwhile, project and studio-based approaches, experiential learning, and creativity-focused visual methods, as illustrated in this paper, could potentially be more effective. Such methods engage students and encourage critical reflection.

Although this paper demonstrated a limited case study that was based on a small sample of student work, the author argues that these methodologies could be scaled up and adopted in disciplines beyond architecture and urban studies. Additionally, by advocating a human-centered, bottom-up understanding of sustainability, the concepts become much more relatable and meaningful to students. ESD courses, regardless of the discipline, should be based on a thorough understanding of sustainability, where students experience and solve problems first-hand and, in the field, rather than through abstract formal presentations or government-sponsored conferences. Further research is needed in this area and in other regional and global contexts.

References

Alawadi K (2017) Rethinking Dubai's urbanism: generating sustainable form-based urban design strategies for an integrated neighborhood. Cities 60:353–366. https://doi.org/10.1016/j.cities.2016.10.012

Alawadi K, Khanal A, Almulla A (2018) Land, urban form, and politics: a study on Dubai's housing landscape and rental affordability. Cities 81:115–130. https://doi.org/10.1016/j.cities.2018.04.001

Al-Naqbi AK, Alshannag Q (2018) The status of education for sustainable development and sustainability knowledge, attitudes, and behaviors of UAE university students. Int J Sustain Higher Educ 19(3):566–588. https://doi.org/10.1108/IJSHE-06-2017-0091

Al-Saidi M (2021) Cooperation or competition? State environmental relations and the SDGs agenda in the Gulf cooperation council (GCC) region. Environ Develop 37:100581. https://doi.org/10.1016/j.envdev.2020.100581

Al-Saidi M (2022) Disentangling the SDGs agenda in the GCC region: priority targets and core areas for environmental action. Front Environ Sci 10:1025337. https://doi.org/10.3389/fenvs.2022.1025337

Al-Saidi M, Zaidan E (2020) Gulf futuristic cities beyond the headlines: understanding the planned cities megatrend. Energy Rep 6:114–121. https://doi.org/10.1016/j.egyr.2020.10.061

Altomonte S (2009) Environmental education for sustainable architecture. Rev Eur Stud 1:82

Bagaeen S (2007) Brand Dubai: the instant city; or the instantly recognizable city. Int Plan Stud 12(2):173–197. https://doi.org/10.1080/13563470701486372

Barhouma M (2023) The path to COP28: UAE climate action pillars. https://epc.ae/en/details/featured/the-path-to-cop28-uae-climate-action-pillars

Benkari N (2013) The "sustainability" paradigm in architectural education in UAE. Proced Soc Behav Sci 102:601–610. https://doi.org/10.1016/j.sbspro.2013.10.777

Bina O, Balula L, Varanda M, Fokdal J (2016) Urban studies and the challenge of embedding sustainability: a review of international master programmers. J Clean Product 137:330–346. https://doi.org/10.1016/j.jclepro.2016.07.034

Boarin P, Martinez-Molina A, Juan-Ferruses I (2020) Understanding students' perception of sustainability in architecture education: a comparison among universities in three different continents. J Clean Product 248:119237. https://doi.org/10.1016/j.jclepro.2019.119237

Crot L (2013) Planning for sustainability in non-democratic polities: the case of Masdar City. Urban Stud 50(13):2809–2825. https://doi.org/10.1177/0042098012474697

Deviren AS, Tabb PJ (2014) The greening of architecture: a critical history and survey of contemporary sustainable architecture and urban design. Ashgate Publishing, Ltd

Elsheshtawy Y (2008a) Cities of sand and fog: Abu Dhabi's global ambitions. In: Elsheshtawy Y (ed) The evolving Arab city: tradition, modernity and urban development. Routledge, New York, pp 258–304

Elsheshtawy Y (2008b) Navigating the spectacle: landscapes of consumption in Dubai. Architect Theory Rev 13(2):164–187. https://doi.org/10.1080/13264820802216544

Elsheshtawy Y (2008c) Transitory sites: mapping Dubai's 'forgotten' urban spaces. Int J Urban Reg Res 32(4):968–988. https://doi.org/10.1111/j.1468-2427.2008.00819.x

Elsheshtawy Y (2009) Dubai: behind an urban spectacle. Routledge, New York

Elsheshtawy Y (2010) Little space, big space: everyday urbanism in Dubai. Sour Brown J World Affairs 17(1):53–71

Elsheshtawy Y (2013) Where the sidewalk ends: informal street corner encounters in Dubai. Cities 31:382–393. https://doi.org/10.1016/j.cities.2012.12.001

Ewers MC, Dicce R (2016) Expatriate labour markets in rapidly globalising cities: reproducing the migrant division of labour in Abu Dhabi and Dubai. J Ethnic Migrat Stud 42(15):2448–2467. https://doi.org/10.1080/1369183X.2016.1175926

Ewers MC, Dicce R (2018) High-skilled migration and the attractiveness of cities. High-Skilled Migrat Drivers Polic 34:176–194

Green LN, Bonollo E (2003) Studio-based teaching: history and advantages in the teaching of design. World Trans Eng Technol Educ 2(2):269–272

Günel G (2020) 8 exporting the spaceship: the connected isolation of Masdar City. In: Molotch H, Ponzini D (eds) The new Arab urban. New York University Press, New York, pp 194–212

Huang L, Wu J, Yan L (2015) Defining and measuring urban sustainability: a review of indicators. Landscape Ecol 30(7):1175–1193. https://doi.org/10.1007/s10980-015-0208-2

Hurlimann A, Cobbinah PB, Bush J, March A (2021) Is climate change in the curriculum? An analysis of Australian urban planning degrees. Environ Educ Res 27(7):970–991. https://doi.org/10.1080/13504622.2020.1836132

Jabareen Y (2011) Teaching sustainability: a multidisciplinary approach. Creat Educ 02(04):388–392. https://doi.org/10.4236/ce.2011.24055

Jabareen Y (2012) Towards a sustainability education framework: challenges, concepts and strategies—the contribution from urban planning perspectives. Sustainability 4(9):2247–2269. https://doi.org/10.3390/su4092247

Kekeya J (2021) Qualitative case study research design: the commonalities and differences between collective, intrinsic and instrumental case studies. Contemp PNG Stud 36:28–37

Khalaf S (2006) The evolution of the Gulf city type, oil, and globalization. In: Fox JW, Mourtada-Sabbah N, Al Mutawa M (eds) Globalization and the gulf. Routledge, New York, pp 254–275

Klopp JM, Petretta DL (2017) The urban sustainable development goal: indicators, complexity and the politics of measuring cities. Cities 63:92–97. https://doi.org/10.1016/j.cities.2016.12.019

Levy JM (2015) Contemporary urban planning. In: Contemporary urban planning. Routledge, New York

Li N, Chan D, Mao Q, Hsu K, Fu Z (2018) Urban sustainability education: challenges and pedagogical experiments. Habitat Int 71:70–80. https://doi.org/10.1016/j.habitatint.2017.11.012

Maruna M (2019) Toward the integration of SDGs in higher planning education: insights from integrated urbanism study program in Belgrade. Sustainability 11(17):4519. https://doi.org/10.3390/su11174519

MoCCAE (2017) National-climate-change-plan-for-the-united-arab-emirates-2017–2050. https://www.moccae.gov.ae/assets/30e58e2e/national-climate-change-plan-for-the-united-arab-emirates-2017-2050.aspx

Molotch H, Ponzini D (2019) The new Arab urban: gulf cities of wealth, ambition, and distress. NYU Press, New York

Moosavi S, Bush J (2021) Embedding sustainability in interdisciplinary pedagogy for planning and design studios. J Plan Educ Res 11:3639

NAAB (2020) NAAB conditions for accreditation, 2020 edn. https://www.naab.org/wp-content/uploads/2020-NAAB-Conditions-for-Accreditation.pdf

National Committee on Sustainable Development Goals (2018) The 2030 agenda for sustainable development United Arab emirates. https://fcsa.gov.ae/en-us/Documents/SDG%20Report%20EN%20Final.pdf

Pacione M (2005) Dubai. Cities 22(3):255–265. https://doi.org/10.1016/j.cities.2005.02.001

Park HY, Licon CV, Sleipness OR (2022) Teaching sustainability in planning and design education: a systematic review of pedagogical approaches. Sustainability 14(15):485. https://doi.org/10.3390/su14159485

Ponzini D (2011) Large scale development projects and star architecture in the absence of democratic politics: the case of Abu Dhabi, UAE. Cities 28(3):251–259

Ponzini D, Alawadi K (2022) Transnational mobilities of the tallest building: origins, mobilization and urban effects of Dubai's Burj Khalifa. Eur Plan Stud 30(1):141–159. https://doi.org/10.1080/09654313.2021.1908233

Prince M (2004) Does active learning work? A review of the research. J Eng Educ 93(3):223–231. https://doi.org/10.1002/j.2168-9830.2004.tb00809.x

QAA, and Advance HE (2021) Education for sustainable development guidance. https://s3.eu-west-2.amazonaws.com/assets.creode.advancehe-document-manager/documents/advance-he/education-for-sustainable-development-guidance_1616774629.pdf

Qamhaieh A, Chakravarty S (2020) Drive-through cities: cars, labor, and exaggerated automobilities in Abu Dhabi. Mobilities 15(6):792–809. https://doi.org/10.1080/17450101.2020.1822103

Ramos S, Vicente P, Passos AM, Costa P, Reis E (2019) Perceptions of the public transport service as a barrier to the adoption of public transport: a qualitative study. Soc Sci 8(5):150. https://doi.org/10.3390/socsci8050150

RIBA Education (2021) RIBA procedures for validation [for UK and international courses in architecture]. https://www.architecture.com/knowledge-and-resources/resources-landing-page/validation-procedures-and-criteria

Rizzo A (2014) Rapid urban development and national master planning in Arab Gulf countries: Qatar as a case study. Cities 39:50–57. https://doi.org/10.1016/j.cities.2014.02.005

Rizzo A (2017) Sustainable urban development and green megaprojects in the Arab states of the gulf region: limitations, covert aims, and unintended outcomes in Doha, Qatar. Int Plan Stud 22(2):85–98. https://doi.org/10.1080/13563475.2016.1182896

Rizzo A (2019) Predatory cities: unravelling the consequences of resource-predatory projects in the global South. Urban Geogr 40(1):1–15. https://doi.org/10.1080/02723638.2018.1505156

Rizzo A, Petruccioli A (2023) Khalifa versus Prometheus: green ethics and the struggle for contemporary sustainable urbanism. Dig Middle East Stud 11:12291. https://doi.org/10.1111/dome.12291

RTA (2021) Dubai the best city for living in the world 2040. The official monthly magazine of Dubai's RTA, pp 4–13. www.rta.ae/links/magazine/masar/Al_Masar_154_Eng.pdf

Schön DA (1984) Toward a marriage of artistry and applied science in the architectural design studio. Sour J Architect Educ 41(4):1425007

Shields JAE (2023) Collage and architecture. Taylor and Francis, Milton Park

Sinakou E, Donche V, Van Petegem P (2022) Action-orientation in education for sustainable development: teachers' interests and instructional practices. J Clean Product 370:133469. https://doi.org/10.1016/J.JCLEPRO.2022.133469

Skirka H (2023) Designs for Dubai's the loop revealed, a sustainable 93km cycling and wellness highway. The National. https://www.thenationalnews.com/travel/2023/02/01/designs-for-dubais-the-loop-revealed-a-sustainable-93km-cycling-and-wellness-highway/

Thomas R, Hsu A, Weinfurter A (2021) Sustainable and inclusive: evaluating urban sustainability indicators' suitability for measuring progress towards SDG 11. Environ Plan B Urban Anal City Sci 48(8):2346–2362. https://doi.org/10.1177/2399808320975404

UNESCO (2020) Education for sustainable development: a roadmap education for sustainable development a roadmap. http://www.unesco.org/open-access/terms-use-ccbysa-en

UN-Habitat (2020) The state of arab cities 2020 financing sustainable urbanization in the arab region. https://unhabitat.org/sites/default/files/2020/06/sacr_2020_executive_summary_en.pdf

Wright J (2003) Introducing sustainability into the architecture curriculum in the United States. Int J Sustain Higher Educ 4(2):100–105. https://doi.org/10.1108/14676370310467131

Yin RK (2018) Case study research and applications, vol 6. Sage Publication, Thousand Oaks, CA

Zaidan E, Abulibdeh A (2021) Master planning and the evolving urban model in the gulf cities: principles, policies, and practices for the transition to sustainable urbanism. Plan Pract Res 36(2):193–215. https://doi.org/10.1080/02697459.2020.1829278

Zaidan E, Al-Saidi M, Hammad SH (2019) Sustainable development in the Arab world: is the gulf cooperation council (GCC) region fit for the challenge? Develop Pract 29(5):670–681. https://doi.org/10.1080/09614524.2019.1628922

Abdellatif Qamhaieh, PhD, is an Associate Professor of Architecture and Urban Studies in the School of Architecture, Arts, and Design at the American University in Dubai.. His current research interests include mobility, urban sustainability, and equity in planning, especially within GCC cities.

Conclusion: Sustainability and Transformation in Arabian Gulf Higher Education

14

Khalaf Al'Abri and Elizabeth Buckner

Abstract

This concluding chapter summarizes the key contributions of the book and points to future directions for research and practice. The chapter argues that collectively, the contributors to this book point out the critical and under-appreciated ways that universities are contributing to sustainability in Gulf societies. The chapter organizes key contributions into three topics: student engagement, governance and research and future visions. In doing so, it invites readers to view higher education in the Arabian Gulf as a field open to new possibilities and, if its energies are fully mobilized, capable of becoming a highly influential force in shaping a more just and sustainable future for the region, its people, and the wider world to which it belongs.

Keywords

Sustainability · Universities · Arab Gulf states · Gulf cooperation council · Transformative pedagogy · Governance · Research

K. Al'Abri (✉)
Sultan Qaboos University, Muscat, Oman
e-mail: kabri@squ.edu.om

E. Buckner
University of Toronto, Toronto, Canada
e-mail: elizabeth.buckner@utoronto.ca

14.1 Higher Education as a Critical Space for Rethinking Sustainability

Economic, social, and urban transformations within the Arabian Gulf states intersect with the growing recognition that the future cannot be postponed. Thus, sustainability has become an existential necessity, an understanding well established in the literature, which treats sustainability not as a narrow technical agenda but as a shift in consciousness and institutional structures (Burns 2012). At this juncture, ambitious national visions coexist with anxieties about climate change, and the aspirations of new generations confront development models that have not been sufficiently tested under environmental and social pressures. Within this context, universities emerge as a critical space for interpreting the present, imagining the future, and reconfiguring relationships among citizens, society, and the state, as well as the environment—responsibilities increasingly reflected in contemporary frameworks for education for sustainable development toward 2030 (Berchin et al. 2021). This role is not just normative or rhetorical but enacted through concrete practices and institutional models within higher education.

Universities hold significance not only for their capacity to produce knowledge, but also for their ability to act as intermediaries between a rapidly transforming world and individuals

K. Al'Abri and E. Buckner (eds.), *Higher Education and Sustainable Development in the Arab Gulf States*, Sustainable Development Goals Series, https://doi.org/10.1007/978-3-032-19559-3_14

seeking to understand change and their place within it. Beyond their individual contributions, the twelve chapters in this book collectively reveal a broad educational landscape shaped by student experiences, faculty practices, institutional structures, national policies, urban transformations, and research agendas. Their key arguments align with critical educational approaches, redefining sustainability not as a narrow technical or environmental agenda, but as a system of meaning, action, and identity. As emphasized throughout this book, sustainability is not merely about emissions reports or energy management plans but rather a way of perceiving the world and reframing fundamental questions: Who decides the shape of the city? Who bears the cost of development? How are resources distributed within society? These questions intersect with global debates about the tangible, direct, and indirect impacts of universities on society, the economy, and the environment, which have been examined both conceptually and empirically (Findler et al. 2019). They also resonate with recent assessments of how Arabian Gulf universities have engaged with the UN's SDGs (Saleh and Adly 2024) and discussions of quality and progress toward 2030 within the Gulf Cooperation Council context (Fathelrahman and AlBreiki 2025).

This book demonstrates that Arabian Gulf universities are no longer merely institutions dedicated to knowledge transmission and credentialing; they have become a space in which national development ambitions intersect with concerns for climate and social justice, as well as a laboratory in which the possibilities of moving from a model of resource-consumption-based prosperity to one of regenerative and protective prosperity are tested. Thus, universities become spaces for rethinking sustainable development rather than merely how it is implemented. This duality of ambition and structural challenge mirrors analyses that position higher education and research in Arabian Gulf states as arenas that reflect both aspiration and institutional constraint (Sellami et al. 2022).

What distinguishes the contributions in this book is that its chapters do not begin by asking how Arabian Gulf states might catch up with global sustainability models. Instead, they pose a more profound question: What does the Arabian Gulf context contribute, in experience and meaning, to the global conversation on sustainability and higher education? From this perspective, Arabian Gulf universities emerge not merely as recipients of knowledge, but as spaces where new approaches take shape at the intersection of rent-based economies in transition, rapidly evolving and diverse cities, and societies negotiating a tension between the local and the global.

Accordingly, rather than simply summarizing developments within Arabian Gulf universities, this book presents a layered perspective suggesting that, rather than being an adjunct to existing structures in the Arabian Gulf, sustainability is a call to reconsider their very foundations. It also positions the discussion within a broader rethinking of higher education globally, framed not as a passive response to change, but as an active means of shaping the future (Jones et al. 2010).

14.2 Students and Transformative Pedagogy

The opening chapters, which examined students' perceptions in the United Arab Emirates, their educational experiences, and their encounters with the city as a space characterized by contradictions of consumption, modernity, and social diversity, reveal that student awareness is neither fixed nor predetermined. Instead, it is formed through ongoing engagement with lived realities—an idea central to the literature on transformative learning, linking consciousness to experience and the re-examination of assumptions (Rodríguez Aboytes and Barth 2020). In the Arabian Gulf, rather than emerging in isolation, student awareness is shaped within rapidly transforming cities where social classes intersect, privilege and vulnerability coexist, and multiple cultures converge within a single urban space. In this book, the city is not treated as a neutral backdrop, but as an educational actor and a mirror through which students come to perceive themselves and others, and to uncover less visible dimensions of dominant development narratives. This approach aligns with scholarship on education for sustainable development that emphasizes the centrality of lived context in shaping learners' critical awareness (Wals 2014).

The chapters demonstrated that theoretical knowledge alone is insufficient to generate a critical understanding of sustainability issues without engagement in lived experiences that compel students to examine urban space, explore power relations, and analyze the relationship between people and places. Here, rather than simply being a built environment or infrastructure, the city is a sociopolitical context through which students learn how the world is shaped, power is constructed, opportunities are distributed, and policies are translated into everyday life. At this point, the empirical analyses presented in this book intersect with the broader literature on transformative education for sustainable development, which emphasizes that learners' awareness shifts when deeply held assumptions are disturbed through exposure to and critical examination of new experiences (Rodríguez Aboytes and Barth 2020).

This perspective highlights a dimension often missing from traditional curricula: as they move beyond the confines of the classroom, students become active subjects capable of linking knowledge to reality, learning to action, and awareness to responsibility. Thus, sustainability originates in individual consciousness before being translated into programs and policies. This vision aligns with global calls for education for sustainable development to transform "ways of thinking" and "ways of being" in the world, rather than merely introducing new content or concepts (UNESCO 2020).

As the book turned to faculty practices, another dynamic emerged that is shaping the future of sustainability in higher education. One Qatari study found that, despite their disciplinary expertise, many academics had a limited, fragmented understanding of education for sustainable development, often reducing sustainability to environmental or technical concerns while overlooking its social, cultural, and political aspects. However, this apparent "deficit" signals not weakness but potential, highlighting a gap that universities can address through systematic professional development that redefines the academic role.

As emphasized throughout these chapters, faculty members are not merely transmitters of content, but makers of meaning, guides for awareness, and designers of the relationships between students and the wider world. In the context of sustainability, this role extends beyond explaining concepts to creating a critical space where students can question their position in the world and recognize that sustainability is not merely a lecture to be delivered, but a relationship built among people, their environment, and society. This understanding aligns with the literature, which argues that embedding education for sustainable development in higher education requires rebuilding institutional cultures and teaching practices rather than simply adding new course units (Leal Filho et al. 2017).

The chapters addressing innovative pedagogies in Kuwait and Dubai, such as project-based learning, urban studios, and fieldwork, offer compelling evidence that pedagogy is not just a tool but the core of education for sustainable development. Sustainability becomes a critical practice when students engage with the city not as a physical entity but as a social structure in which class relations are reproduced, identities intersect, and consumption patterns are formed. Thus, when students are asked to analyze a neighborhood, observe urban spaces, or trace mobility patterns, they encounter previously unseen dimensions of the world. The city appears as a dynamic map reflecting state policies, market practices, migration flows, and power relations. Thus, education becomes a gateway to understanding the world, not merely a mechanism for knowledge retention. This notion aligns with contemporary perspectives on universities' role in redefining what it means to be human (Molotch and Ponzini 2019).

Such transformative pedagogy does not just add an "activity" to a course; it reshapes the course itself. It transforms students from passive recipients into partners in knowledge production, the city from context into curriculum, and sustainability from an abstract concept into lived experience. This approach aligns directly with the core insights from the transformative learning literature on sustainability (Rodríguez Aboytes and Barth 2020). From this perspective, education for sustainable development in the Arabian Gulf has little meaning unless it is critical, experiential, and unconstrained by the closed classroom, consistent with global scholarship in this field.

14.3 Governance and Research

The chapters on governance and national policies, particularly the Omani experience, suggest that the primary challenge is not the lack of vision but rather weak mechanisms for translating vision into action. Arabian Gulf states have announced ambitious strategies aligned with the UN's SDGs, often supported by political commitment and funding. However, universities often face a gap between national visions and their capacity to implement them. Rather than minor administrative shortcomings, this gap reflects structural complexity arising from multiple actors, structured decision-making, and unclear divisions of responsibility among government ministries, universities, and regulatory bodies. These characteristics have been documented in recent scholarly analyses of higher education and research governance in the Arabian Gulf (Sellami et al. 2022) and in case studies examining institutional transformation in specific Arabian Gulf contexts (Michaleczek and Sellami 2025).

However, this book presents this gap as a foundational opportunity to rebuild governance within universities, strengthen institutional autonomy, empower sustainability-focused units, and redefine the relationship between universities and the state as a partnership rather than a purely executive arrangement. In the Arabian Gulf, achieving sustainability will require more than just slogans or strategy documents (UNESCO 2020); it will require institutional structures that embed sustainability into the university's "DNA," including budgeting processes, decision-making systems, leadership criteria, and performance metrics. This notion aligns with frameworks that conceptualize universities' sustainability impact as institutional and long-term, rather than as a collection of isolated initiatives (Findler et al. 2019).

The chapters focusing on research, particularly at Sultan Qaboos University, have also demonstrated that research has moved beyond being a purely decorative academic activity to become a knowledge-practice foundation for climate response in the Arabian Gulf. Universities with strong research infrastructures can act as partners in formulating national policies rather than merely serving as educational institutions. These chapters highlighted how research centers generate data on water, energy, urban development, and climate change, and translate these findings into policies, recommendations, interventions, and even new understandings of the university's role. Thus, research becomes not merely a means of understanding but also a means of action, decision-making, and imagination.

As this book portrays, Arabian Gulf universities stand at the threshold of becoming a state's "knowledge brain" in environmental, climate, and urban matters, provided they are recognized as policy partners rather than mere implementers of predefined programs. This recognition requires building stronger institutional bridges between research outputs and national strategic planning bodies, ensuring that university data and analyses become integral to the decision-making process rather than optional. It also emphasizes the need to expand the concept of "quality" in higher education beyond internal efficiency or competitive attractiveness to include universities' contributions to achieving the UN's SDGs—a notion increasingly reflected in recent scholarly analyses in the Arabian Gulf (Fathelrahman and AlBreiki 2025).

The chapters focusing on Arabian Gulf cities, particularly Dubai, reveal an additional dimension of the sustainability project. Here, the city is not a neutral context, but a social text through which ways of living are interpreted, inequalities are tested, and development models are constructed. The city reveals sharp contradictions: architectural modernity alongside parallel social strata;

urban affluence intersecting with the vulnerability of marginalized groups; and official narratives of smart and sustainable amenities contrasting with uneven access to public spaces, services, and environmental and social security. These contradictions position the city as a pedagogical platform, highlighting that sustainability is not just environmental or technical but a fundamentally social project that requires scrutinizing urban organization and the social relations on which cities are built. This perspective resonates with critical literature on Arabian Gulf cities as spaces where wealth, ambition, and imbalance coexist (Molotch and Ponzini 2019).

By perceiving the city as an educational context, students, researchers, and faculty members can understand how policies shape everyday life; how work, housing, and mobility patterns are produced; how class differences intersect with environmental practices; and how sustainability becomes a question of justice and representation before it becomes a matter of policy. Through this urban lens, the university becomes part of the urban fabric rather than an isolated island—a partner in reimagining the city itself, not merely its own.

14.4 Ethical Dimensions and Future Visions

The impact of this book extends beyond offering recommendations or documenting case studies. It reconstructs the intellectual discourse on sustainability in the Arabian Gulf and repositions the university from a consumer of existing models to a producer of new ones. Its epistemic contribution lies in decentering knowledge production, framing the Arabian Gulf not merely as an object of study but as a source of theory and conceptual innovation. Arabian Gulf universities operate within complex contexts that can generate new approaches to understanding relationships between people and cities, states and markets, and demographic diversity and urban sustainability.

At the institutional level, this book provides a framework for universities to redesign their internal structures to integrate sustainability into policies, curricula, research, and governance, rather than treating it as a parallel or temporary project. This transformation represents not an administrative improvement but a fundamental redefinition of the university's function and societal role—a notion captured in studies framing universities' sustainability impact as multidimensional and long-term (Findler et al. 2019).

Politically, this book offers conceptual tools to help decision-makers understand the gaps between vision and implementation and reconsider the roles universities can play in shaping national climate commitments (UNESCO 2020). Socially and culturally, this book revisits questions of justice in the Arabian Gulf, foregrounding long-neglected issues such as the right to the city, resource distribution, representation of marginalized groups, and the cultivation of student awareness that recognizes social diversity as integral to sustainability rather than an obstacle to it.

More broadly, this book suggests that Arabian Gulf universities are moving toward a future in which they function as platforms for design rather than execution alone—bringing together knowledge, research, policy, and practice. The city will increasingly be seen not as an external context but as part of the curriculum, transforming Arabian Gulf cities—with their demographic diversity and contradictory practices—into living laboratories that link theory and reality and reveal the everyday meaning of sustainability.

Within this context, the ethical role of higher education becomes more pronounced. As the chapters illustrated, sustainability is not merely an environmental commitment or a set of operational plans but a project of cultivating individuals capable of imagining alternative futures, questioning structures that produce inequality, and recognizing that sustainable living is a relationship not only with nature but also with others. Thus, sustainability emerges as an inherently ethical educational project.

Tracing the chapters and their analyses reveals that this book goes beyond merely describing current realities; it opens horizons for envisioning what Arabian Gulf universities might become. The accumulated insights demonstrate that sustainability in the Arabian Gulf does not simply borrow ready-made models or adhere to a single global framework; instead, it is continually reshaped through the region's social, economic, and political realities.

These visions are not presented as closed theories or claims to be a complete explanatory framework but rather as forward-looking interpretations grounded in the opportunities, challenges, and structural tensions revealed across the chapters. They offer a more nuanced understanding of the role of Arabian Gulf universities in a world accelerating toward environmental crises, deepening social justice awareness, and transforming human relationships with cities, public space, work, and knowledge. In light of this book, the future Arabian Gulf university can be imagined as a hub for post-oil imaginaries, where alternative development scenarios are tested; as an urban laboratory for environmental justice, where the city becomes a site of reflection and intervention rather than mere habitation; and as a bridge between knowledge and policy, where research becomes embedded within decision-making structures rather than remaining peripheral.

These visions are not offered as a ready-made roadmap, but as an open invitation to universities, policymakers, researchers, and students to rethink their roles and responsibilities. They serve as a reminder that the university is not a fixed destiny, but an institution in constant formation; that sustainability is not a stable endpoint, but an ongoing process of review, reconstruction, and negotiation among competing interests and uneven aspirations.

Thus, this concluding chapter does not offer finality, but rather a framework for thinking beyond this book. It invites readers to view higher education in the Arabian Gulf as a field open to new possibilities and, if its energies are fully mobilized, capable of becoming a highly influential force in shaping a more just and sustainable future for the region, its people, and the wider world to which it belongs.

References

Berchin II, de Aguiar Dutra AR, Guerra JB (2021) How do higher education institutions promote sustainable development? a literature review. Sustain Develop 29(6):1204–1222. https://doi.org/10.1002/sd.2219

Burns TR (2012) The sustainability revolution: a societal paradigm shift. Sustainability 4(6):1118–1134. https://doi.org/10.3390/su4061118

Fathelrahman EM, AlBreiki GAA (2025) Quality of higher education in the Gulf cooperation council countries: a review of progress toward the 2030 SDGs. J Adult Contin Educ 31(1):96–108

Findler F, Schönherr N, Lozano R, Reider D, Martinuzzi A (2019) The impacts of higher education institutions on sustainable development: a review and conceptualization. Int J Sustain High Educ 20(1):23–38. https://doi.org/10.1108/IJSHE-07-2017-0114

Jones P, Selby D, Sterling S (eds) (2010) Sustainability education: perspectives and practice across higher education. Routledge

Leal Filho W, Brandli L, Castro P, Newman J (eds) (2017) Handbook of theory and practice of sustainable development in higher education (vol 1). Springer

Michaleczek IJD, Sellami A (2025) Higher education and sustainability in the GCC countries: the case of Qatar. In: Zaidan E, Tok E, Al-Fadala A, Cochrane L, Zaidan E (eds) The future of education policy in the State of Qatar (Gulf Studies, vol 21, pp 367–387). Springer. https://doi.org/10.1007/978-981-97-9667-0_19

Molotch H, Ponzini D (eds) (2019) The new Arab urban: Gulf cities of wealth, ambition, and distress. New York University Press

Rodríguez Aboytes JG, Barth M (2020) Transformative learning in the field of sustainability: a systematic literature review (1999–2019). Int J Sustain High Educ 21(5):993–1013. https://doi.org/10.1108/IJSHE-05-2019-0168

Saleh SA, Adly HM (2024) Measuring the impact of higher education in promoting sustainable development goals: analysis of Arab universities' performance. Problemy Ekorozwoju 19(1):261–274

Sellami A, Arar K, Sawalhi R (2022) Higher education and scientific research in the Arabian Gulf states: opportunities, aspirations, and challenges. Routledge

UNESCO (2020) Education for sustainable development: a roadmap. UNESCO

Wals AE (2014) Sustainability in higher education in the context of the UN Decade of education for sustainable development: a review of learning and institutionalization processes. J Clean Prod 62:8–15. https://doi.org/10.1016/j.jclepro.2013.06.007

Khalaf Al'Abri is an Associate Professor of Education Systems and Policies at Sultan Qaboos University,

Oman, where he also serves as Head of the Department of Educational Foundations and Administration. His research focuses on higher education policy, governance, sustainability, and internationalization in the Gulf region. He has led several national and international research projects on sustainable development, quality assurance, and global citizenship education. Dr. Al-Abri has published extensively on university rankings and higher education reform. He is President of the Gulf Comparative Education Society and collaborates with UNESCO and other international partners to promote education for sustainable development.

Elizabeth Buckner is Associate Professor of Higher Education at the Ontario Institute for Studies in Education (OISE) at the University of Toronto, where she also holds the Tier 2 Canada Research Chair in Higher Education for Sustainable Global Development. Her research examines how global trends affect higher education and one strand of her work focuses on how universities are integrating sustainable development into their missions and operations.

GPSR Compliance

The European Union's (EU) General Product Safety Regulation (GPSR) is a set of rules that requires consumer products to be safe and our obligations to ensure this.

If you have any concerns about our products, you can contact us on ProductSafety@springernature.com

In case Publisher is established outside the EU, the EU authorized representative is:

Springer Nature Customer Service Center GmbH
Europaplatz 3
69115 Heidelberg, Germany

Batch number: 10731334

Printed by Printforce, the Netherlands